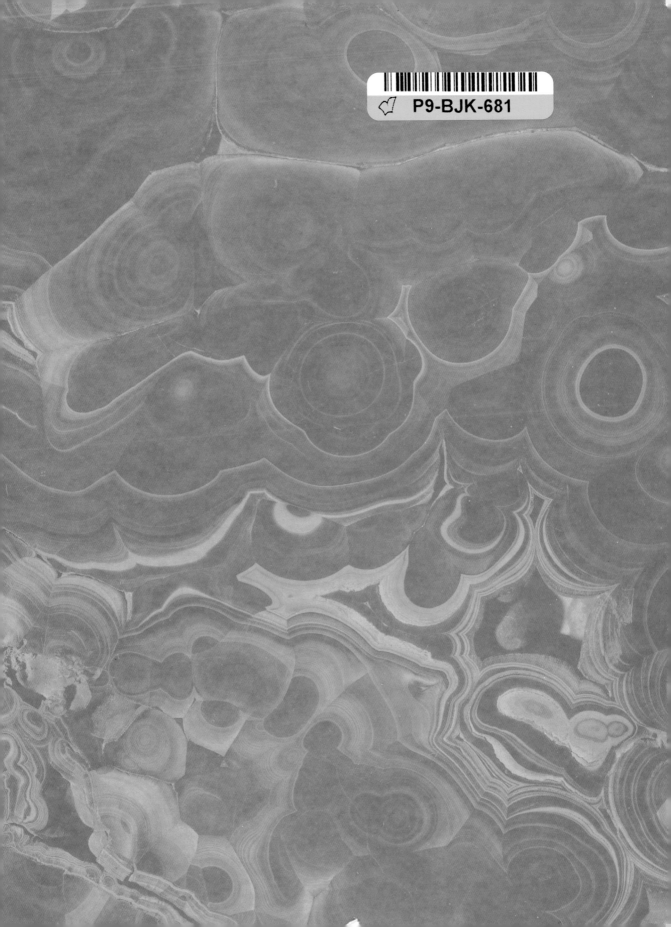

GEMS
OF THE
WORLD

CALLY OLDERSHAW

FIREFLY BOOKS

A FIREFLY BOOK

Published by Firefly Books Ltd. 2008

First printing

Publisher Cataloging-in-Publication Data (U.S.)

Oldershaw, Cally.
      Gems of the world / Cally Oldershaw.
[256] p. : col. photos., ill., maps ;   cm.
Includes index.
Summary: Guide to the identification and use of gemstones. Includes the geology, chemistry and properties of gems; the diamond industry, what to look for when buying gemstones, and how to care for gems.
ISBN-13: 978-1-55407-367-2
ISBN-10: 1-55407-367-7
1. Precious stones — Popular works. I. Title.
553.8 dc22   QE392.O4347 2008

Library and Archives Canada Cataloguing in Publication

Oldershaw, Cally
      Gems of the world / Cally Oldershaw.
ISBN-13: 978-1-55407-367-2
ISBN-10: 1-55407-367-7
      1. Precious stones. I. Title.
QE392.O385 2008   553.8   C2007-905873-6

Published in the United States by
Firefly Books (U.S.) Inc.
P.O. Box 1338, Ellicott Station
Buffalo, New York 14205

Published in Canada by
Firefly Books Ltd.
66 Leek Crescent
Richmond Hill, Ontario L4B 1H1

Cover design: Erin R. Holmes

Printed in China

CONTENTS

We use the word "gem" in everyday language, for example "she is a real gem" and "this book is a little gem." In this context a gem is something special, highly valued and well-thought of, something to be treasured, with special attributes.

Gemstones are also treasures. Their unique qualities have been valued throughout the ages, across continents and by different peoples, from our earliest ancestors to the present-day. It may have been the color or the crystal shape of a gemstone, or a brightly colored shell that first attracted the attention of someone who then bent down to pick it up. Something special about it would have encouraged that person to keep it, to own it, maybe to put it in a special place such as a bag hung around the neck, for safekeeping, to polish or make a hole in it, or to tie it on to clothing as an adornment or as a piece of jewelry.

Gems and jewels are associated with the rich and famous. We may admire the jewels worn by our favorite film star, celebrity, or sportsperson. We may even aspire to own some particularly fine piece ourselves. Gems have been worn as a symbol of status, adorning the crowns of royalty – a visual reminder of wealth, success and achievement to both the wearer and the observer. The power and energy ascribed to certain gemstones are an attribute defined by mystics and healers. The tales of famous stones, the luck they may hold or the curse they may inflict, can captivate an audience.

In choosing this book, you may already have been captivated by the "specialness" of gemstones, or you may be interested in knowing more about them, you may work with gemstones, for example, as a scientist, a student, a designer, or a jeweler. Whether you are a scientist, an artist, or just interested, we hope that this book inspires and informs you. It is intended as a guide to the beautiful and fascinating world of gemstones. It shows you the glorious diversity of colors and the incredible crystal shapes of these wonders of the natural world. There is information about the optical and physical properties of the gemstones, where and how they are found, how they can be worked and how they should be cared for.

▼ Long prismatic purple crystals of amethyst.

CRYSTALS AND GEMSTONES

Some crystals look as fragile as glass and are incredibly rare, but they have an inherent strength. Crystals may take millions of years to form, or may form as you watch. They may have been formed in rocks deep beneath the Earth's surface, or they may be survivors of mountain-building episodes or devastating volcanic eruptions, or they may have been washed into rivers and streams to be retrieved maybe millions of years after their formation. These survivors are nature's treat: perfect and brightly colored crystals formed in dark, deep rocks.

Generally speaking, gemstones are minerals that have formed as sufficiently clear, large crystals that can be cut and polished for use as pieces for personal adornment or *objects d'art* such as sculptures, inlays, and so on. Pre-cut gemstones and minerals in matrix are also collectable. In addition to the mineral gemstones there are also other materials that can be used for adornment, such as pearl, shell, amber and other derivatives of plants or animals. These are called organic gems.

◀ *Crystal fragment of citrine.*

WORKING WITH GEMSTONES

But for a gemologist (someone who studies the science of gemstones, their physical and optical properties and their origins) or a jeweler, what are the special attributes of gemstones? For gemstones to be used in jewelry, ideally they should have three main attributes: beauty, durability, and rarity. Beauty and rarity have a direct impact on the value of a gemstone, the more beautiful and rare the greater the value. The color and clarity of a gemstone are just two of the aspects that a gemologist or jeweler will take into account when studying a gemstone or choosing the best gem material for jewelry. Durability, the strength of the gemstone, will affect how it can be worked, cut and set or mounted, and how it should be cared for to avoid it becoming scratched, cracked, or otherwise damaged.

▶ *Citrine gems.*

However, not all gemstones possess all three. For example, some may be insufficiently durable to use as a cut gemstone in a ring, but may be good as a piece within a brooch, protected from damage by the mounting. Some materials are best suited for fashioning as beads or cut *en cabochon* (with a domed surface), others look their best when faceted (fashioned with a number of flat, polished surfaces).

The skill of the lapidary, jeweler or jewelry designer lies in their ability to recognize the qualities of the gem materials, working with them to their best abilities to produce a piece that is both admired and sought after. They will need to find the best compromise between what is possible and what is practical. Jewelers also need to understand the strengths and weaknesses of various gem materials in order that they can be confident of the identity of the material they are trading and also so that they can advise the customer on the best care and cleaning methods, and most importantly disclose any treatments, for example oiling or heating, which may affect the value, use, or durability of the article.

◀ *Citrine cross pendant necklace.*

IMITATIONS, FAKES, AND FORGERIES

Not all gemstones are what they seem. A gemstone that has similar properties to a more valuable or rare specimen may be used to imitate it. Color can be misleading: for example, the color of a red spinel might be mistaken for a ruby. Glass, plastic and other materials both natural and manmade can also be used to imitate gemstones.

Even the assumption that a gemstone has been formed naturally may not be true. Synthetic gemstones have the same chemical and physical properties as their natural equivalent, but they are made in the laboratory. Part of the excitement of being a gemologist is to know how to use your eyes and the various pieces of equipment available in order to distinguish the imitations, fakes, and forgeries from the real gems.

G ems are minerals that form in rocks and sediments. Each mineral or mineral group has a precise chemical composition, made of the same elements or chemical ingredients. These ingredients are given in its chemical formula – the chemical code that defines the mineral and the elements from which it is made.

GROWTH AND COLOR

In addition to the main elements, there may be other elements that do not appear in the chemical composition; these accessory elements may affect the properties of the mineral, such as the color. For instance, the corundum family, which includes ruby and sapphire, has a very simple chemical composition (Al_2O_3) – two atoms of aluminum (Al) joined with three atoms of oxygen (O). Pure corundum would be colorless, but nature seldom enables pure or perfect gemstones to form, so corundum is usually colored. Chromium and vanadium are accessory elements (trace elements) that give ruby its red color, iron and titanium give sapphire its blue color, and other combinations give rise to the many other colors of sapphire such as green, yellow, and pink.

Changes in the surrounding temperature or pressure may affect the concentrations and composition of the fluids in which some gemstones are formed. Accessory elements may be available throughout the formation of the gemstone or at intervals dependent upon the prevailing environmental conditions.

The variation in elements available during formation may result in a gemstone having an irregular growth or color pattern, color banding or variations in the concentration of color throughout the gemstone. Color banding associated with formation may be obvious, for example in the curved banding of agates. Concentrations of color, in growth zones or straight bands, can sometimes be seen in gemstones such as rubies, sapphires, and amethyst.

The formation of a gemstone may be interrupted, growth may be sporadic, and there may even be intervals during which the gem deteriorates, or is partly dissolved, before growth continues. A gemstone (the host) may contain crystals of a different mineral within it (the inclusions). These inclusions may be formed before, after, or at the same

▲ *Rutilated quartz.*

▶ *Geodes, or weathered-out cavity linings, are lined with well shaped crystals (left). Mineral veins yield many good gem specimens (right).*

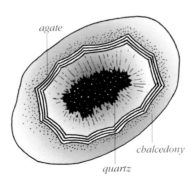

agate

chalcedony

quartz

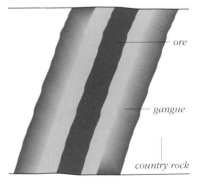

ore

gangue

country rock

time as the host. They may be seen as flaws that detract from the beauty of the gemstone or features that actually enhance the gem, as is the case for rutilated quartz, where the inclusions of rutile within the quartz may add to the rarity, desirability, and value.

GEM QUALITY

Rocks are made of minerals, but not all minerals form crystals that are large enough or in a good enough condition, or clear enough of cracks and unattractive inclusions, to be used as gemstones. The vast majority of crystals are too small to be seen, even with a hand lens (or loupe) with x10 magnification. Of the larger crystals, only a very small proportion will be of sufficient quality to be cut and fashioned as gems; these are generally referred to as "cuttable" or "gem quality" crystals.

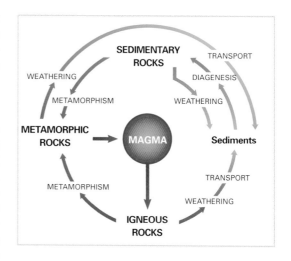

▲ *Rock cycle diagram.*

ROCK TYPE

Gemstones form in the three main rock types: igneous, sedimentary, and metamorphic rocks. The formation of rocks is a continuous cycle of events (see rock cycle diagram). The rock cycle follows the changes through which rock progresses from its initial formation, being broken down into smaller fragments by wind, rain, and snow (weathering), transported down rivers and streams (erosion), and finally deposited in riverbed sediments and the sea (deposition). These sediments may then form new rock types and the cycle continues.

IGNEOUS

Igneous rocks are those that are formed as a result of volcanic or magmatic activity. Beneath the surface of the Earth, molten rock (magma) rises and falls. As molten rock rises, there are changes in pressure and temperature that affect mineral formation. If the molten rock reaches the surface, it may be erupted as lava and other volcanic rocks. These volcanic igneous rocks are often referred to as extrusive igneous rocks. As the lava cools down, minerals form within the rock. Generally, the slower the rate of cooling, the larger the crystals that form.

Peridot is commonly formed in basalt lava, an example of an extrusive igneous rock. Other gems that crystallize from lavas as they cool include zircon, ruby and sapphire, moonstone, topaz, and red beryl.

The largest and the best specimens of peridot have been found on the Red Sea island of Zerbirget, Egypt. Zerbirget was formed as the result of a slab of mantle rock (part of a layer from deep beneath the surface of the Earth), having been thrust up on to the crust rather than being erupted from a volcano. Diamonds also form deep underground (at depths of up to 650 feet/200 meters) within the mantle and have been brought to the surface in volcanic eruptions that were far larger than any present-day eruptions.

▲ *Kimberlite, one of only two principal sources of diamonds.*

▲ *Tourmaline-bearing pegmatite rock.*

Where the magma is not erupted, but does rise sufficiently for the magma to cool, crystals may form. These large, slow-cooling magma masses are the origin of some of the largest gem crystals found. The igneous rocks formed in this way are called intrusive igneous rocks. Granite is a common intrusive igneous rock.

Pegmatites are intrusive igneous rocks that produce a greater range of gemstones than any other rock type and also some of the largest, record-breaking gemstones. The large crystals are formed as the water-rich portion of a granite-like molten rock is put under increased pressure as it is squeezed into fractures in surrounding rock. As the molten rock begins to solidify, the elements that it contains begin to form crystals. Initially, the outer part of the molten rock will solidify. The largest gemstones and some quite rare gem varieties may form gem pockets at the center of the pegmatite from the hot concentrated mineral-rich fluid, which is the last to crystallize.

Gems formed in pegmatites include topaz, tourmaline, kunzite, and members of the beryl family such as the blue aquamarine and the pink morganite. Pegmatites occur around the world, but the largest gem producers are the mines of Minas Gerais, Brazil. Other pegmatite areas include the Pala area of San Diego County, California, the Nuristan area of northeast Afghanistan, the Sverdlovsk (Yekaterinburg) region of the southern Urals, Russia, and the Altai Mountains of northwestern China.

Gems also grow from hydrothermal fluids (from *hydro* meaning "water" and *thermal* meaning "hot") that escape from magmas and may contain rare elements such as fluorine and beryllium. As the hydrothermal fluids move away from the magma along fractures and fissures within the surrounding rock, they may solidify and fill them, forming mineral veins. Close to the surface, hydrothermal veins may also include elements carried by groundwater and other near-surface waters. Amethyst, topaz, benitoite, and emerald are some examples of gemstones found in hydrothermal veins.

SEDIMENTARY

When rocks are weathered and eroded (see rock cycle diagram, page 7), they break down into boulders, pebbles, smaller rock fragments, and finally to sands or muds. During this process, the gemstones they

Mudstone

Fossiliferous freshwater limestone

Orthoquartzite

deposited in rivers and the ocean. The gemstones may become concentrated in gravels, cemented or compressed to form sedimentary rocks.

With sufficient time and the right conditions of temperature and pressure, water can dissolve rocks. The minerals of the rocks may be transported elsewhere and on encountering other elements, may react with them and produce new minerals or may be deposited directly from solutions. These form, for example, in veins or cavities within rocks or as crusts on surfaces of rocks or other materials.

New minerals may also form as deposits when the water-levels change, water cools, or where the water evaporates and leaves a mineral residue. Examples of gemstones formed in sedimentary rocks as a result of the evaporation, cooling or transportation of mineral-rich fluids include turquoise, malachite, rhodochrosite, agate, and amethyst.

Sedimentary rocks may contain fossils. The organic remains may decompose to leave minerals within the rock or leave cavities into which minerals may be transported. For example, opal forms from a silica gel within a sedimentary rock. Large specimens, such as fossilized dinosaur bones, may be replaced by opal and small irregular voids within a rock can be filled.

METAMORPHIC

All rocks, including sedimentary and igneous rocks, may be altered by pressure and/or temperature to form metamorphic rock (*meta* means "changed," *morph* means "shape"). Metamorphic rocks are often associated with large, mountain-building episodes and enormous temperatures and pressures, which fold and fault large masses of rock. The mineral composition of the rock alters as the temperature and pressure changes, and new metamorphic minerals form.

The area affected by metamorphism may cover only a few inches or feet, usually as a result of local faulting or folding, or hundreds of miles due to larger regional tectonic events such as mountain building. For example, the rubies of Burma (Myanmar) are associated with the mountain-building episode that formed the Himalayas more than 65 million years ago. Jadeite is an example of a gemstone formed by metamorphism as a result of high pressures.

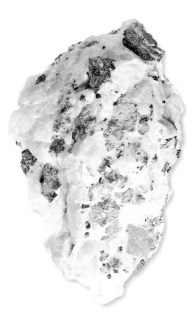

▲ *Diopside-garnet marble, an example of a metamorphic rock.*

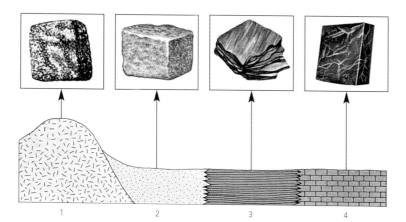

◀ *Metamorphic rock. Some common metamorphic rocks are shown below: igneous rock (1) may become gneiss or schist; sandstone (2) may become quartzite; shale (3) may become slate or, if the metamorphism is more pronounced, gneiss; limestone (4) may become marble.*

9

A study of gemstone mining and retrieval covers every mining method, from the traditional searches in streams and rivers using just a pan or sieve, to the ultra-high technology and research used in diamond mines deep underground. Ultimately, any source of gemstones will only be mined or exploited if the source is viable. It must be possible to mine or retrieve the gems at a profit to make the venture a viable business.

ALLUVIAL DEPOSITS

The oldest and most traditional methods are still in existence in areas where gemstones are near the surface, relatively easy to find and retrieve, and labor is cheap. For example, in Indonesia, Malaysia, Sri Lanka, and India, local inhabitants search rivers and streams, as well as the gravels and sediments that were once ancient riverbeds or streams. They may use basic equipment such as baskets, pans or buckets to retrieve river gravels and sediments, and a sieve or pan to begin to separate the gemstones.

This is possible because the gemstones are generally heavier than the surrounding mud, pebbles or rock fragments. As a pan of water and sediment is "jiggled," the gemstones settle in the pan as the lighter constituents and the water are washed over the pan's edge. The heavier concentrate may be sieved to separate the larger gems, or spread out on tables or cloths to be hand-sorted and the gems found by eye.

▼ *Traditional panning for alluvial diamonds in Sierra Leone.*

These gemstone localities, found associated with the sedimentary rocks of rivers and streams, are called "alluvial deposits" or placer deposits. They are secondary deposits: they are not found in the rock in which they were formed, but where they have been transported to as a result of weathering and erosion.

Gemstones that survive the journey tend to be those that are sufficiently hard to withstand the conditions without breaking rather than those that are heavily included, that fracture easily, or break along cleavage planes. They are generally harder and heavier than surrounding minerals and in water they tend to sink faster and are therefore not carried as far. The surviving gemstones will generally become concentrated in pockets or areas along the riverbanks or within sediments, as gem gravels.

Alluvial deposits, such as the gem gravels of Sri Lanka and Burma (Myanmar), contain a diverse range of gemstones including ruby, sapphire, spinel, chrysoberyl, topaz, tourmaline, and garnet. Gem gravels may contain good crystals, such as diamond, ruby, and spinel, but generally the gemstones show signs of wear and tear. Crystals may be fragmented and rounded, and the surface may be scratched or frosted due to friction (rather like frosted glass). Because nature has already sorted out the weaker specimens, the percentage of gem-quality stones within gem gravels is high, making their retrieval worthwhile. In fact, more gemstones are retrieved from gem gravels than from any other type of deposit.

Because a number of gemstones that are associated with gem gravels are often found together, the discovery of one type of gemstone from a gem "association" (sometimes called a tracer gem) can be used by exploration teams and prospectors to "home in on" or "trace" potential gemstone mining areas. Another technique is to map the courses of ancient river beds or present-day rivers and streams, and follow tracer gems downstream in the hope of finding areas where the gemstones are in sufficiently large enough concentrations to be retrieved.

In Colombia, emeralds form within thin layers of white limestone in soft, black carbonaceous shales. Enormous trucks are used to transport the soft shales to washing plants, where the emeralds are retrieved by separating the harder limestone from the soft shale. Landscapes are altered as hillsides are removed. In addition to the large-scale removal of the rocks, there may also be groups of people working on a much smaller scale. As the remaining shales are weathered, the gems may be washed out and transported downhill into the valleys below. Local people search the river beds and sediments for emeralds, often just using spades or their bare hands.

Where the rock is harder, these methods are insufficient. For example, rubies can be found in the metamorphic rocks in which they have formed, as well as being found as alluvial deposits. In some instances, picks and drills may be sufficient to prise loose the gemstones from the parent rock (the host), while in others the metamorphic rock has to be mined, crushed, washed, and sorted to retrieve the gemstones.

▲ *Bucket wheel dredge moving overburden, Namaqualand, South Africa.*

DIAMONDS

Primary deposits

Diamonds are mined from diamond-bearing rocks on a larger scale and with more highly mechanized methods than any other gemstone, because of their value and range of uses. The process is highly computerized and every stage is carefully monitored to ensure the highest standards of safety and security.

The diamond pipes are mined from the surface (surface mining, open-pit mining or open-cast mining), removing the loose overburden using enormous hydraulic shovels and trucks, and transporting it and the rock beneath to processing plants to be crushed and washed before removing the diamonds. As the pipe is dug out, a large pit is formed. Once a pit is about 1,000 feet (300 meters) deep and it is no longer possible to surface mine, underground tunnels and shafts are needed to excavate and remove the diamond-bearing rock (underground mining).

Rock is taken in ore cars (open box-like containers moving along a track) to an underground crusher (or sizer) that breaks the ore down to pieces less than 6 inches (15 cm) across. The ore is then taken by conveyor belt to skips that lift the ore to the surface. Overground crushers usually break the ore down to even smaller pieces in the range of 1–2 inches (2.5–5.0 cm). Larger diamonds that are not noticed prior to the second crusher will be shattered, so there is an incentive to keep a look out for loose diamonds before they reach the crusher.

The crushed ore is then processed to concentrate the heavier minerals and separate the diamonds. Using circular rotary washing pans the lighter material rises toward the top of the muddy water mixture and overflows. The heavier material which contains the diamonds settles to the bottom and is drawn off. This concentrate may undergo further crushing and separation processes before the diamonds are removed using an X-ray fluorescence separator or less commonly a grease table. Diamonds have an affinity to grease and stick to a grease table (a conveyor belt smothered with a layer of grease). They are then scraped off using a hot knife. Diamonds emit light (fluorescence) when subjected to an X-ray beam. As the diamonds fall from the feed belt, the fluorescence triggers a jet of air which pushes them so that they fall into the diamond bin rather than the waste bin. The final sorting is done by hand.

The better quality stones may be fashioned as cut gems, while even the non-gem-quality diamond can be used as an abrasive or for other industrial purposes. The percentage of gem-quality to industrial-quality diamond varies from mine to mine and from continent to continent. Pipe mining of igneous rock (kimberlite or lamproite) produces a greater total yield of diamonds than alluvial deposits, but typically only 20–25% are of gem quality. Pipe mining produces gem-quality cuttable diamonds and those that are used for industrial purposes (industrial diamonds). The total diamond production and the proportion of gem-quality cuttable diamonds compared with industrial diamonds varies from mine to mine. As 50% is lost during the cutting and polishing of a diamond, a 2-carat piece of rough material is needed to produce a 1-carat polished diamond. Statistics from De Beers diamond mines have been used to estimate that, on average, for every polished diamond of 1 carat (0.2 g), 250 tons of rock will have had to be mined.

▼ *Diamond-bearing rock is transported to rock crushers.*

Secondary deposits – alluvial and marine

The highest concentrations of gem-quality diamonds are found in alluvial deposits (river, beach, and seabed sands and gravels), where diamonds have been concentrated by weathering and erosion. Around 80% of diamonds recovered from alluvial deposits are of gem quality because most of the flawed diamonds have been broken up during transport.

Diamonds found in the diamond-bearing coastal sands on the shores and seabed of the Namibian coast have been transported down rivers to the beach and sea. Early reports of the "sparkling coastal sands" were due to the presence of diamonds on the surface which could be collected by the opportunistic traveler. Nowadays, the mining process is highly mechanized. Sand barriers hold the tides back while sand is transported in trucks to sorting plants and huge bucket wheel dredges scoop up large quantities of sand which are sifted for diamonds. Once the overburden of sand has been removed, miners use brushes to check the hollows in the underlying rock, and hand-picking any diamonds that may have accumulated there. The used sand is replaced and the sand barriers breached to allow the return of the tides.

Special ships have been developed to retrieve diamonds from the sands of the seafloor. The ships follow a straight course, beneath which is a hose that sucks up water, sand and diamonds. The sand is sieved and sorted on board and the diamonds collected in cans. The cans are sealed awaiting collection by helicopter.

▲ *Marine diamonds are retrieved from the sea using ships that vacuum the sand off the seabeds.*

FROM THE MINE TO THE MARKET

There are a number of routes by which gemstones reach the customer. They may be mined industrially on a large scale, with a high degree of mechanization and security, owned by international organizations with infrastructure ensuring a defined route to the market. At the other extreme they may be retrieved locally by artisans working in family or village groups hand-picking gemstones from alluvial deposits. Security may be lacking or left to those with weapons. There may be no formal routes for the gemstones, they may be traded in the mines, in local towns or cutting centers, or smuggled or exported to be cut and fashioned elsewhere. They may even be used to support drugs barons or guerrilla warfare in areas of conflict.

The majority of gemstones follow a prescribed route from the mine to the market. They are sold by miners and mine owners to experienced gem buyers. A gem buyer will generally concentrate on a particular region of the world or a particular gem material, for example a ruby or sapphire buyer, a jade buyer or an expert in gems and crystals from Brazil. They assess the material and pass it to cutters and polishers. The gems then follow the route to wholesalers, retailers and the consumer.

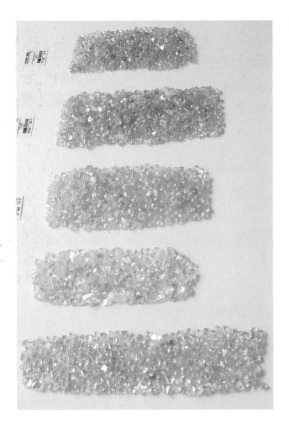

▲ *Rough diamonds are sorted by size, shape, color, and clarity at the Diamond Trading Company.*

The stability of the diamond industry has been due in part to the way in which a system of controls was set up in the last century, initially with the consolidation of holdings by Cecil John Rhodes in Kimberley in South Africa and later by Sir Ernest Oppenheimer. Sir Ernest Oppenheimer, as Chairman of De Beers Consolidated Mines, steered the consolidation of diamond mines and the formation of the Diamond Producers' Association and the Diamond Trading Company (DTC) in 1934, which together made up the Central Selling Organization (CSO).

The CSO was able to offer stability to producers and customers by holding back diamonds and maintaining stock piles when production outstripped demand, and releasing them to support production when demand was greater than production. By consolidating mines, more productive mines could support those in early stages of production or those nearing the end of their productive life, while ensuring that supplies were maintained. At one point De Beers produced over 90% of the world's diamonds and was the only major supplier of diamonds, marketing their diamonds through the CSO.

The Diamond Trading Company, the sales and marketing arm of De Beers, still markets almost 45% of the world's production of rough diamonds by value, approximately 130 million carats with a value of about US$13 billion (2006). The process by which this is achieved is via the route referred to as "the diamond pipeline" (see pages 16–17).

AGGREGATION AND SORTING

London used to be the center for all DTC operations, including aggregation, sorting, sales, and the location of De Beers' Head Office. Recent changes have taken place in the diamond pipeline following negotiations with a number of African countries and others.

The main sorting operation of De Beers has been in London since 1934. De Beers also has major sorting operations in Kimberley (South Africa), Windhoek (Namibia) and Lucerne (Switzerland). Botswana is the single largest source of rough gem diamonds for De Beers and for the world market, producing some 30% of world totals.

The practice has been for South Africa's diamonds to be sorted and sold at viewings abroad. Recent agreements will enable a percentage of local diamonds to be made available for resale to local cutters and producers. In 2006, representatives from 18 African countries formed the African Diamond Producers Association (ADPA). The Association aims to bring together representatives of the African countries that produce diamonds, as well as the big diamond companies, in order to agree policies to support diamond mining and the African diamond industry.

As part of moving the sorting operation from London to Gaborone (Botswana) by 2009, a new company, the Diamond Trading Company Botswana, has been established. In 2006, the building which was built to process a maximum of 14 million carats per year, had to deal with record production of 32 million carats. In order to accommodate the new systems, plans include building a larger sorting building, installation of new and improved machines and sorting stations, and training for the increased number of employees needed. This move is part of De Beers' "beneficiation" program which aims to provide more benefit to local communities, using domestically mined diamonds to generate jobs and add value to the rough diamonds in the county of origin.

There are other changes at De Beers; it is divesting itself of some mines, for example in Tanzania where production is small, and reducing activity in alluvial ventures. The DTC buys only about US$400 million worth of alluvial diamonds and is not involved with the alluvial mines of, for example, Brazil. The DTC is selling many of the small mines of South Africa and even the famous mines are undergoing change. The Kimberley mines are more or less closed and diamond production exhausted, and the Cullinan Mine (formerly named the Premier Mine) is being sold.

▲ *A DTC client views rough diamonds during "site week."*

BUYING ROUGH DIAMONDS

As a result of recent agreements, compilation of site boxes also takes place in Botswana. Some parcels (sites) are sold locally to cutting operations in South Africa, Namibia and Botswana. In South Africa, the arrangements with the DTC have been passed as law, while in Botswana and Namibia agreements have been by negotiation with the DTC. In Botswana aggregated stock can be sold to clients locally, though not every diamond may be suitable for particular buyers. In Namibia the Government has asked for some purely Namibian goods to be made available for sale in Namibia. In South Africa a proportion of the South African production is sold by the South African Government through their state diamond traders.

DTC site holders are encouraged to support the shift away from the more traditional centers and to support the programs for sustainable development that encourage local involvement and "beneficiation." Although aggregation will be moving away from London and some sales will be held internationally, the major sales will still be held in London every five weeks (10 times a year).

▼ *Brown diamonds. Not as rare as other colors such as pink and blue, brown diamonds are often described as cognac or champagne in color.*

THE DIAMOND PIPELINE

I. AGGREGATION

The proportion of gem-quality cuttable diamonds compared with industrial diamonds varies from mine to mine as does the tendency to have larger diamonds, a particular shape or color of diamond. The first stage in the diamond pipeline is to collect the rough diamonds from all the mines that use the DTC route and mix them together (aggregation) locally or in London. Around 65% of the world's diamonds are from Africa, mainly Botswana, South Africa, Namibia and Tanzania.

Some aggregation takes place in the country of origin, for example by DTC International in Namibia and South Africa (and Canada by the end of 2007 or early 2008).

2. SORTING

The aggregate is then sorted into categories. Aggregation can benefit the client who buys the diamonds as it mixes like for like from a number of sources, boosts each category and allows more diamonds to be offered to clients requesting particular attributes such as size, shape, or color.

A small part of the sorting process is mechanized but the skills and expertise of trained sorters are most important. Dia-

▲ *Aerial view of the De Beers' Koffiefontein diamond mine, situated southeast of Kimberley, North Cape Province, South Africa. Open-cast mining has ceased, and mining is now carried out underground.*

▶ *Rough diamonds being sorted at the Diamond Trading Company in London.*

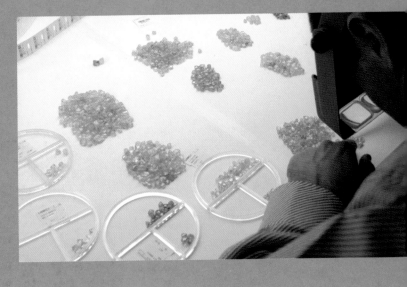

monds that are not of sufficient size (carat), clarity, or color to be cut as gemstones are separated for industrial use (termed industrial diamonds). The remaining cuttable or gem-quality diamonds are sorted into as many as 16,000 different categories depending on a combination of size and shape, color, and clarity.

3. BUYING ROUGH DIAMONDS

Industrial-quality diamonds are sold at a fixed price per carat. Gem-quality diamonds are divided into parcels (sites) to be stockpiled or sold at one of 10 sales that are held by De Beers each year. There are fewer than 200 regular buyers or "site holders," mainly large dealers

who market both rough and polished diamonds. Most of the sales take place in London. There are also sales in Johannesburg (South Africa) and Lucerne (Switzerland).

The site holder places an order, stipulating the range of diamonds preferred. The range may be to satisfy the needs of certain cutting centers, for example larger-sized diamonds are preferred by the US, small- and medium-sized diamonds are preferred by India. The allocation is decided by the DTC and the site holder has no choice about the contents. The site holder is invited to view the parcel, agree the valuation of the contents, and decide whether or not to purchase the entire parcel.

4. CUTTING AND POLISHING

The world's main cutting and trading centers are Antwerp, Mumbai, New York, Johannesburg and Tel Aviv (see map, pages 20–21). Site holders assess their purchase, the diamonds may resold as rough material, or cut and polished before selling via a network of private offices, diamond clubs and the 24 registered diamond trading centers (diamond bourses) in the major diamond centers of the world.

5. WHOLESALE TO RETAIL

Diamond traders, designers and manufacturers in the wholesale business sell to retailers and other outlets from which customers buy their gemstones or jewelry.

▲ *Diamond dealers assess the value of diamonds and trade both rough and cut material at bourses such as the diamond bourse at Antwerp.*

▼ *Rough diamonds are either sold on or taken to a cutting center.*

RUSSIA

In 2006, Alrosa, Russia's diamond monopoly, sold approximately US$23 million worth of rough diamonds at the 18th international diamond auction for special-sized diamonds hosted by the Russian Diamond Chamber in Moscow rather than through the DTC diamond pipeline. There were more than 100 parcels, containing a total of 794 diamonds weighing around 13,300 carats, including 15 diamonds weighing more than 50 carats each. The biggest diamond offered weighed 125.72 carats. Representatives of 43 companies from Russia, India, Israel, Belgium, China, Japan, Switzerland and the US bid at the auction.

Russia will try to maintain its share of the world rough diamond market at its current level of slightly more than 25% for the next 10 years. By the end of 2009, De Beers will have stopped buying rough diamonds from Russia (at present about US$2 billion worth a year).

CANADA

By the end of 2009, diamonds from the Canadian diamond-producing mines of Victor (Ontario) and Snap Lake (186 miles/300 km north of Yellowknife) will be included in the aggregation. Diamonds from Canadian mines owned by Rio Tinto and BHP Billiton do not use the DTC to market their production, worth more than US$3 billion in 2006.

AUSTRALIA

Australia's diamond production is not marketed through the DTC diamond pipeline. Rio Tinto, the company that owns the Argyle Mine, famous for its pink diamonds, has an annual auction of pink diamonds. According to Rio Tinto, the Argyle mine produces an average of 30 million carats of small, colored diamonds per year. In 2006, all 65 pink diamonds from the annual production were sold to an exclusive group of 26 invited bidders. The diamond collection included a wide variety of polished stones in a broad range of pink hues and a small selection of violet diamonds, with sizes ranging from 0.49 carats to 2.03 carats. Their bids were at an all-time high, exceeding those of each annual tender held since 1984.

Details of the winning bids and total revenue generated from the tender are not disclosed in order to protect bidders' confidentiality; however, pink diamonds are known to command prices exceeding US$400,000 per carat, 20 times the price of equivalent white (colorless) diamonds, due to their rarity and market demand. A news report stated that a Brisbane jeweler had bought a red diamond at the auction, which he says is "one of only seven ever found in the world." The owner said he was "prepared to look at offers over US$2 million" and has already had three offers "over seven figures." The brilliant-cut diamond named "The Lady in Red" has a diameter of 5.13 mm and weighs 0.54 carats.

▼ *Pink diamonds. Found in Australia's Argyle Diamond Mine and the Williamson Mine, Tanzania. Pink and blue diamonds are rare and consequently command a higher price than other fancy-colored diamonds such as yellow or brown.*

CUTTING AND POLISHING

Antwerp (Belgium) is still a major cutting center, though Thailand and India are rapidly increasing their capabilities. In the past as much as 80% of the total world production of rough diamonds was thought to pass through the diamond center of Antwerp. With the expansion of other diamond trading centers, this figure is likely to drop. Dubai is trying to establish itself as a diamond trading center. Africa is also gaining experience, and cutting centers in South Africa, Namibia, and Botswana are likely to expand.

INDIA

India employs a vast number of people in cutting and polishing factories; it has been estimated that of every 12 polished diamonds, 11 have been polished in India. Many Indian diamond traders still have an office in Antwerp, but as the Indian trade increases and the global industry changes, they may move their offices elsewhere.

▲ *Jeweler's workshop in the diamond district of Mumbai, India.*

CHINA

China is investing a great deal in training and gaining a reputation as a diamond-manufacturing center. China's largest diamond mine produces about 100,000 carats a year which, in the global arena, is still small; however, there are plans to expand the industry. There are already around 100 polishing factories, employing more than 20,000 people. According to unofficial statistics provided by China's diamond industry, annual sales of consumer-grade diamonds are currently growing at a rate of 40%, making China one of the world's fastest-growing diamond jewelry markets.

Wuzhou, a small city in Guangxi Province, is China's largest costume jewelry center. It is also China's biggest gemstone trading center (colored gemstones and diamonds). The Wuzhou Gemstone City site covers an area of 5.4 acres (40,000 square meters), and has more than 500 shops and separate businesses.

WHOLESALE TO RETAIL

Diamonds may be resold as rough material or cut and polished before selling via a network of private offices, diamond clubs and the 26 registered diamond trading centers (diamond bourses) in the major diamond centers of the world including London, Antwerp, Tel Aviv, New York, Mumbai (Bombay), Singapore, and Bangkok. Each diamond may be involved in a number of transactions before being purchased by a wholesaler, retailer or private individual. Internet and shopping channel sales continue to increase; for example, one internet site has listings of over US$2.8 billion worth of diamonds every day. Diamonds, however, do not tend to be traded once they reach the customer. The marketing of diamonds centers on its value as an emotional gift that the owner will not wish to part with.

DIAMOND CUTTING AND TRADING CENTERS

The major diamond manufacturing centers include Israel, Belgium (Antwerp), India (Mumbai and Surat), and New York; cutting and polishing also takes place in South Africa, Botswana, Russia, China, Sri Lanka, Thailand, Vietnam, and Mauritius.

DIAMOND BOURSES AROUND THE WORLD

There are 26 diamond trading centers (diamond bourses) registered as members of the World Federation of Diamond Bourses (see *Useful Links* on the facing page).

Map key
- ■ Major diamond manufacturing centers
- • Diamond bourses

■ In **New York**, most of the offices are located in what is known as the Diamond District – between Fifth Avenue and the Avenue of the Americas – from 46th to 48th Streets. New York is considered the "gateway" to the US market and is also a manufacturing center for large diamonds.

■ **London** has been the center for the distribution of rough diamonds for more than 300 years. The Diamond Trading Company (DTC) offices (the sales and marketing arm of De Beers) are here. Site holders, both manufacturers and dealers in rough, meet in London, once every five weeks, 10 times a year, to inspect their diamond allocations.

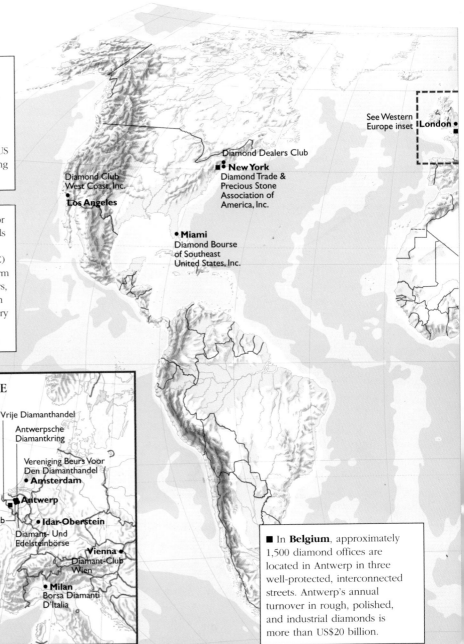

See Western
Europe inset | London •

Diamond Dealers Club

■ • **New York**
Diamond Trade &
Precious Stone
Association of
America, Inc.

Diamond Club
West Coast, Inc.
• **Los Angeles**

• **Miami**
Diamond Bourse
of Southeast
United States, Inc.

WESTERN EUROPE

Vrije Diamanthandel

Antwerpsche
Diamantkring

Vereniging Beurs Voor
Den Diamanthandel
• **Amsterdam**

London •
The London Diamond
Bourse and Club

■ **Antwerp**

Diamantclub
Van
Antwerpen

Diamant- Und
Edelsteinbörse
• **Idar-Oberstein**

Vienna •
Diamant-Club
Wien

• **Milan**
Borsa Diamanti
D'Italia

■ In **Belgium**, approximately 1,500 diamond offices are located in Antwerp in three well-protected, interconnected streets. Antwerp's annual turnover in rough, polished, and industrial diamonds is more than US$20 billion.

The World Federation of Diamond Bourses (WFDB) was founded in 1947 to unite and provide bourses trading in rough and polished diamonds and precious stones, with a common set of trading practices. The WFDB provides a legal framework and convenes to enact regulations for its 26 member diamond bourses. New branches for the Bourse include Kaliningrad, Yatutsk and Yekaterinburg (2007). The World Diamond Council, set up by the World Federation, works with more than 35 governments, the European Union and the United Nations in a bid to stop the trade in conflict diamonds.

USEFUL LINKS
World Federation of Diamond Bourses
http://www.worldfed.com/website/
World Diamond Council
http://www.worlddiamondcouncil.com
Diamond High Council (HRD)
http://www.diamonds.be

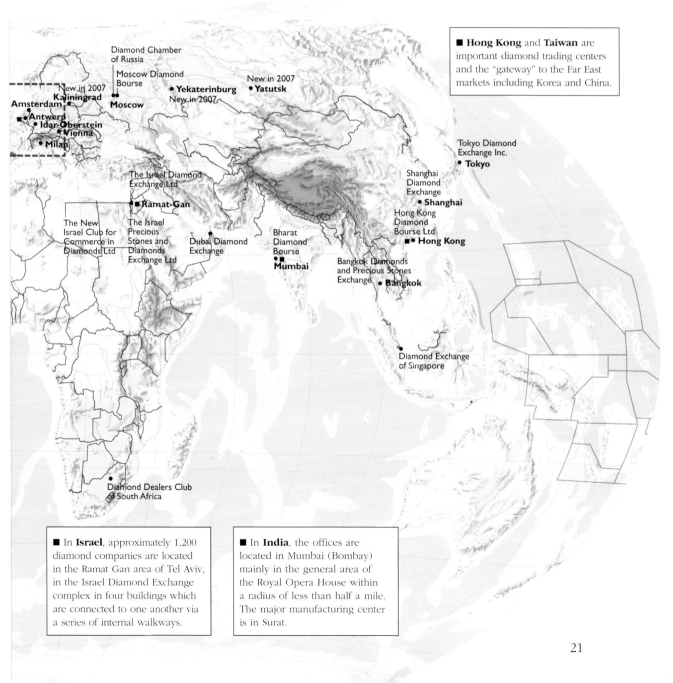

■ **Hong Kong** and **Taiwan** are important diamond trading centers and the "gateway" to the Far East markets including Korea and China.

■ In **Israel**, approximately 1,200 diamond companies are located in the Ramat Gan area of Tel Aviv, in the Israel Diamond Exchange complex in four buildings which are connected to one another via a series of internal walkways.

■ In **India**, the offices are located in Mumbai (Bombay) mainly in the general area of the Royal Opera House within a radius of less than half a mile. The major manufacturing center is in Surat.

Greater knowledge of the diamond industry has highlighted areas where the industry has been unable to stop unlawful or unethical behavior. Probably the best-known example is that of conflict diamonds: diamonds that serve to finance illegal behavior, guerrilla warfare and wars in areas of conflict. The release of the Hollywood film *Blood Diamond*, starring Leonardo DiCaprio as a South African mercenary pursuing a rare pink diamond through Sierra Leone's diamond-fueled civil war, has served to bring the issue to the general public.

About 75,000 people were killed in the West African conflict, which ended in 2002. The diamond industry, acutely aware of the negative impact that the film might have on diamond sales, has already begun an awareness campaign citing the benefits to developing countries of legally traded diamonds, which contribute about US$8.4 billion to African economies annually. Around 65% of the world's diamonds are from Africa, mainly Botswana, South Africa, Namibia and Tanzania.

Prior to the release of the film, and in order to combat conflict diamonds, the diamond industry had introduced a certification scheme, the Kimberley Process, which aims to track the route of a diamond from the mine to the market, whether it follows the DTC diamond pipeline or not. The certification scheme applies only to natural unworked (rough) diamonds and not to cut and polished stones.

THE KIMBERLEY PROCESS

1. The Kimberley Process is an import/export control regime designed to stop conflict (blood) diamonds entering legitimate trade.

2. Diamond-producing countries control the production and transport of rough diamonds from the mine to the point of export.

3. Diamonds are sealed in a transparent, tamper-proof, security bag by a government official. The official logs the export registration number, carat weight and export value (US dollars) on a "Certificate of Export Origin." The matching counterfoil is sent to the destination country.

4. Customs officers at the importing country check the authenticity of the counterfoil and check details with the exporting officials. The shipment is not dispatched until the counterfoil and details have been verified.

THE KIMBERLEY PROCESS CERTIFICATION SCHEME

The Kimberley Process Certification Scheme (KPCS, Kimberley Process or KP) is an international governmental certification scheme that was set up following three years of negotiation between the governments of major diamond trading and producing countries, representatives of the diamond industry and non-governmental organizations (NGOs) to prevent the trade in diamonds that fund conflict.

The 70 member countries, civil society and the diamond industry developed the certification scheme governing trade in over 99% of the world's rough diamonds.

Countries that agree to participate must pass legislation to enforce the Kimberley Process. They must also set up control systems for the import and export of rough diamonds. Participants are only allowed to trade rough diamonds with other participants. The aim is to prevent conflict (blood) diamonds from entering the Kimberley Process system and to maintain confidence in the diamond marketing process.

The Kimberley Process participants (governments) and observers (the diamond industry, NGOs) meet once a year to discuss the implementation of the scheme. In 2006 about 300 delegates, representing the diamond industry, producer countries and non-governmental organizations (NGOs) from 70 countries, gathered for the annual review of the trade, worth about US$37.6 billion a year.

Since the advent of the process 99.8% of diamonds traded have been certified as sourced from conflict-free areas. Although the scheme makes it more difficult for diamonds from areas of conflict and rebel-held areas to reach international markets, there are still significant weaknesses in the scheme. The scheme is voluntary and there are no sanctions or fines for those who are found to be contravening the guidelines. Stronger controls (including government controls) and better systems for identifying suspicious shipments of diamonds through international trading centers are just two of the proposals put forward to increase the efficiency of the system.

Other weaknesses include the fact that there is no one official standardized "Certificate of Export Origin" (Kimberley Process Certificate), which makes it difficult for customs officers to verify the authenticity of the certificates.

The Kimberley Process covers only rough material and is a voluntary scheme. However, the diamond industry has also signed up to a code of conduct and a system of warranties and guarantees that all diamonds (including rough, cut, and polished stones) have been purchased from a legitimate source. However, this again is a voluntary and self-regulated scheme open to abuse, without checks or sanctions.

Weaknesses in the Kimberley Process are found across the diamond pipeline, including in countries with trading, cutting and polishing centers. A recent United States Government Accountability Office (GAO) report shows that conflict diamonds may be entering the US because of major weaknesses in the implementation of the Clean Diamond Trade Act, the US law which implements the Kimberley Process Certification Scheme.

In the UK, implementation of the Kimberley Process is centered on the Government Diamond Office (GDO), a team of officers operating out of the Foreign and Commonwealth Office. The GDO is responsible for authorizing exports of diamonds by issuing the Kimberley Process Certificate.

To have most impact on preventing the trade in conflict diamonds, government monitoring and verification of industry compliance with the Kimberley Process must explicitly be made a minimum requirement of the scheme, and statistics on the flow of diamonds between participating countries should be published.

▲ *South African mercenaries help to transport diamonds safely out of Sierra Leone.*

Tektite, a form of silica-rich glass, is amorphous

Diamond crystal: octahedron

Pyrite crystal: pyritohedron

Chrysoberyl: twinned crystal

Each gemstone group or variety has a unique chemical composition that describes its essential elements or ingredients. Each gemstone is made up of atoms, the smallest building blocks of the structure. Where the atoms are constituted in a regular and repeating three-dimensional pattern, they are said to possess crystal structure and are referred to as crystalline.

AMORPHOUS

Where there is no crystal structure and atoms are randomly arranged, the material is said to be amorphous (from *amoph* meaning "without shape" or "without form"). Obsidian (volcanic glass) is an example of an amorphous material that can be used as a gemstone.

CRYSTALLINE

Most natural and artificial gemstones are crystalline. Crystalline materials have directional properties, both physical and optical, that are as a result of the three-dimensional arrangement of atoms and the type of bonds between them. As the gemstone forms, it grows by adding layers of atoms. The many ways that the atoms can combine is reflected in the many different crystal shapes including those with flat faces and sharp edges, needle-like points, or with smooth and rounded surfaces.

Although the internal crystal structure of the crystalline gemstone may define its external shape (habit), this is not always the case. Conditions during formation may be such that an irregular shaped crystal forms, or a crystal may have been broken, cut or polished. Whatever the final shape of the crystal, the internal structure is still the same, it is still crystalline.

TWINNING

The crystal structure may have parts that are reflected, repeated incorrectly or rotated, which may result in the creation of a twin crystal. At the junction of the parts of the twin, the change in orientation of the crystal structure (lattice) and direction of growth may be seen as a re-entrant angle at the surface. Repeated twinning (lamellar twinning) can sometimes be seen as layers in gemstones. Twinning of a crystal structure may be due to changes in temperature or pressure during or after formation.

Corundum: trigonal spindle-shaped form

Rutile: tetragonal form

Spinel (Ceylonite): cubic form

Beryl (heliodor): hexagonal prism

POLYMORPH

Occasionally the chemical composition is such that it can form more than one crystal structure. Each structure is called a polymorph (from *poly* meaning "many" or "more than one," and *morph* meaning "shape"). Which of the polymorphs forms will depend upon factors such as conditions during formation (for instance, temperature or pressure). Diamond and graphite, for example, are formed from the element carbon, but the carbon atoms have bonded in different ways, forming different crystal structures in each of the minerals.

Diamond: octahedron with curved faces

ISOMORPH

Some gemstones have a range of chemical compositions, but have the same crystal structure. They are called isomorphs. An example is the garnet group. At opposite ends of the range are the two end members and between these are the members of the isomorphous series. Almandine and pyrope are respectively the iron-rich (Fe) and the magnesium-rich (Mg) end members of the garnet family.

Almandine crystal

CRYPTOCRYSTALLINE

Crystals composed of many crystal structures that have grown together are termed polycrystalline. Where the crystal structure is too small to be seen with the eye and a microscope is needed, the crystals are termed microcrystalline or cryptocrystalline (from the Greek *crypto* meaning "hidden").

CRYSTAL SYMMETRY

Crystals can be classified into groups depending upon their symmetry, which in turn is defined by the degree of regularity in the arrangement of the atoms in the crystal structure. This arrangement also affects their optical and physical properties, for example how they react to light, the way they look and feel, and their hardness.

The main method by which a gemologist usually defines the symmetry of a crystal is by reference to its planes of symmetry and axes of symmetry (sometimes referred to as mirror symmetry and rotational symmetry).

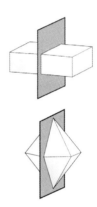

Planes of symmetry

PLANE OF SYMMETRY

A plane of symmetry is an imaginary plane (or mirror) that divides a crystal such that the image on one side of the plane is the mirror image of that on the other side.

AXIS OF SYMMETRY

An axis of symmetry is an imaginary line (or thread) that runs through a crystal and about which the crystal can be rotated, in such a way that it looks the same two, three, four, or six times during a complete circle (rotation) of 360 degrees.

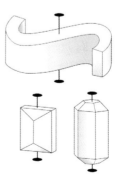

Axes of symmetry

REFERENCE AXES OF THE CRYSTAL SYSTEMS
AND SOME EXAMPLES OF EACH

Cubic

Tetragonal

Orthorhombic

Monoclinic

Triclinic

Hexagonal and trigonal

CRYSTALLOGRAPHIC AXIS (REFERENCE AXIS)

A crystal is also defined by its reference or crystallographic axes. The crystallographic axis is an imaginary line that runs through a crystal and indicates both the direction and length of the repeating pattern of the atoms (the lattice structure). The lengths of the crystallographic axes are proportional to the repeat directions in the three principal directions (a, b, and c) in the crystal structure or lattice (x, y, and z). Crystallographic axes are not the same as the axes of symmetry, although they may run parallel to each other.

CRYSTAL SYSTEMS

There are seven crystal systems based on their crystal symmetry and crystallographic axes (length and direction) and all crystal structures can be assigned to one of these systems.

Each of the crystal systems, except triclinic, has horizontal axes (a and b) and a vertical crystallographic axis (c axis). The direction and length of each axis and the angles between them (alpha, beta, and gamma) are defined by the crystal structure (lattice).

Note: In the United States and Canada, it is more usual to refer to six crystal systems. This is because trigonal and hexagonal are usually grouped together (as hexagonal).

SYSTEM	SYMMETRY	REFERENCE AXES
CUBIC	Four three-fold axes.	Three axes mutually at right-angles, and of equal length.
TETRAGONAL	One vertical four-fold axis.	Three axes mutually at right-angles; one axis conventionally held vertically, differing in length from the other two.
ORTHORHOMBIC	Either one two-fold axis at the intersection of two mutually perpendicular planes, or three mutually perpendicular two-fold axes.	Three axes mutually at right-angles, all of different length.
MONOCLINIC	One two-fold axis.	Three axes of unequal length; two axes are not at right-angles; the third, the symmetry axis, is at right-angles to the plane containing the other two.
TRICLINIC	Either a center of symmetry or no symmetry.	Three axes, all of unequal length, none at right-angles to the plane containing the other two.
HEXAGONAL	One vertical six-fold axis.	Four axes, three of equal length, arranged in a horizontal plane; the fourth perpendicular to this plane and of different length from the other three.
TRIGONAL	One vertical three-fold axis.	Same as for hexagonal.

OPTICAL PROPERTIES

THE IMPORTANCE OF LIGHT

Without light, we would not be able to appreciate the bright green of emerald, the rich red of ruby, the soothing blue of sapphire, the sunshine yellow of heliodor, or the sunset pinkish orange of padparadscha. It is the energy of light that brings the colors of the gemstones to life.

Color is the most obvious optical property, but the way that light interacts with the structure of a gemstone also gives it other optical properties which combine to make each gemstone unique. Effects produced by light reflecting off a gem's surface, off layers or structures within or passing through a gemstone assist the gemologist in identification, but also give each gemstone its particular quality and beauty.

Light reflected off the surface of a gemstone gives the gemstone its luster, whether it looks metallic, glassy, or dull. The brightness and quality of the luster depends on the condition of the surface, the degree of polish, and the refractive index (see explanation on the next page). The greater the amount of light that is reflected back toward the eye, both from the surface of a gemstone and having entered the stone, from internal reflections such as off the inside of the back of the stone (pavilion facets), the brighter the stone will appear. The amount of light, which can pass through a gemstone, will define whether it is transparent, translucent (enough light for an image to be seen through the gemstone, but not enough, for example, for writing to be seen clearly), or opaque.

▼ *Table and examples of the different types of luster in gemstones.*

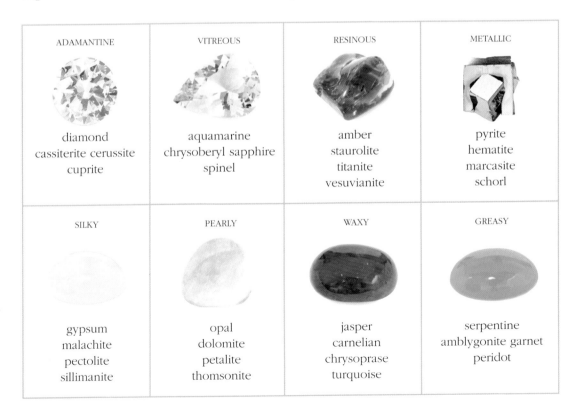

ADAMANTINE	VITREOUS	RESINOUS	METALLIC
diamond cassiterite cerussite cuprite	aquamarine chrysoberyl sapphire spinel	amber staurolite titanite vesuvianite	pyrite hematite marcasite schorl

SILKY	PEARLY	WAXY	GREASY
gypsum malachite pectolite sillimanite	opal dolomite petalite thomsonite	jasper carnelian chrysoprase turquoise	serpentine amblygonite garnet peridot

Dispersion

Sunlight (white light) is made up of the colors of the rainbow (spectral colors). Each color corresponds to a different wavelength of light and a different energy, and each is refracted (bent) to a different degree (measured as its refractive index) as sunlight shines through the rain, effectively separating and spreading the colors to form the rainbow (dispersion). In a gemstone such as a diamond, which has a high dispersion, flashes of rainbow colors appear to come out of the stone as the gem or its light source is moved.

The colors of the rainbow are the colors of the visible spectrum, the wavelengths of light with the energies we are able to see. On either side of the visible spectrum other energies such as X-rays, ultraviolet rays and infrared rays interact with the crystal structure to modify the color or produce optical effects such as fluorescence and phosphorescence.

Refractive index, birefringence, and the refractometer

Light that enters a gemstone is refracted (bent) from its original path in air as it enters the more dense medium. Cubic and non-crystalline materials are singly refractive, which means that the light is refracted equally in all directions. Gemstones from the other crystal systems are doubly refractive. In doubly refractive gemstones, the light entering the gem is split into two rays of light; each is slowed and refracted by a different amount. Where the difference is large, such as in calcite, double refraction can be seen as a double image through the gemstone. In sphalerite, a doubling of the image of the back facets (pavilion facets) can be seen when viewing them through the front of the stone (crown facets).

There is a mathematical relationship between the angle at which light strikes a gemstone and the angle of refraction, from which the refractive index (RI) of a gemstone can be calculated. The refractive index of most gems can be measured accurately using a refractometer or Brewster Angle Meter and can be used to help identify the stone. Singly refractive gemstones have a single refractive index. Doubly refractive gemstones have a range of refractive indices; the difference between the maximum and minimum values is the birefringence (DR).

REFLECTION EFFECTS

Interference – iridescence and play of color

Rainbow effects such as those seen in cleavage cracks and the iridescence of labradorite feldspar and hematite are due to interference as light is reflected by thin layers (films) within the gemstone. In moonstone feldspar, the effect is known as schiller (sheen), adularescence or opalescence and the iridescent colors on the surface of the pearl is called the "orient of pearl." As light is reflected, there is

▲ *Dispersion in a mixed-cut diamond.*

▲ *Round brilliant-cut sphalerite gem.*

▼ *Brewster Angle Meter.*

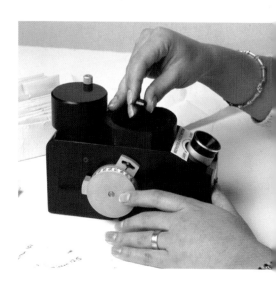

▲ *Labradorite cut as a cabochon.*

Opal

Cat's-eye chrysoberyl

Star sapphire

interference of the wavelengths. Where wavelengths coincide, the color corresponding to the wavelength may be enhanced, in other places waves may cancel each other out and that color will no longer be seen.

In opal, interference of light occurs as it passes between the regularly arranged spheres that make up its structure. The size of the spheres and the distance between them affects the amount of dispersion and resultant play of color; as well as the direction from which the opal is viewed. Small spheres produce only the blues and violets as the opal is turned, while large regularly packed spheres show the full range of rainbow colors.

CHATOYANCY, ASTERISM, AND SILK

Other internal reflection effects include chatoyancy, asterism and silk, which are caused by inclusions. Chatoyancy is the cat's-eye effect seen on some gemstones that are cut as a cabochon (polished as a rounded dome) and best seen under a bright light such as a spotlight or torchlight. Light is reflected off a parallel arrangement of elongated or acicular (needle-like) crystal inclusions of minerals such as rutile or tourmaline, fibers or long tube-like cavities. Examples of gemstones that may show a cat's-eye include quartz, beryl, ruby, sapphire, and tourmaline.

Asterism forms star stones where there are two or more sets of parallel inclusions instead of a single set. Stars may have four, six, 12, or even 24 arms (rays). Star sapphires and rubies generally have six arms, which lie parallel to the crystallographic axes. Other gemstones that may show stars when cut as cabochon include garnet, quartz and spinel.

Where the inclusions or cavities are not present in large enough concentrations to form a star, they may be seen as silk, with the light reflected from patches of parallel inclusions, such as can often be seen in sapphires.

COLORED GEMSTONES

The color of a gem largely depends on how it absorbs or reflects light energy. When white light strikes a gem, some of the spectral colors are absorbed in preference to others (preferential absorption). Those that are not absorbed pass through the gem or are reflected back, giving the gem its color.

Some gemstones have a characteristic absorption spectrum that, with the use of a gemological instrument called a spectroscope, can be used to assist identification or, for example, to distinguish between two gemstones with a similar appearance but different spectra, such as ruby and garnet. Through the spectroscope, the absorption spectrum of the gemstone looks like an incomplete rainbow, with black lines or bands replacing some of the color. The lines and bands signify the energies with wavelengths that correspond to those colors that have been absorbed; the remaining energies give the gem its color.

IDIOCHROMATIC AND ALLOCHROMATIC GEMSTONES

Where the color of a gemstone is caused by elements that are an essential part of the chemical composition, the gem is termed idiochromatic (from *idio* meaning "same" and *chroma* meaning "color"); for example, the green color of peridot is due to iron (Fe), an essential part of its composition (FeS_2).

Most gemstones are allochromatic (from *allo* meaning "other") and are colored by small amounts of other elements or impurities called trace elements that are not an essential part of the chemical composition. The most common trace elements are the metals chromium, vanadium, iron, titanium, copper, and manganese. The bright green of emerald and demantoid garnet is caused by chromium, which also gives the bright red of ruby. Sapphires are colored blue by iron and titanium, or green, yellow or brown by iron. Manganese colors spessartine-garnet orange and rhodonite pink. Heating, irradiation, and other means of altering, enhancing or destroying color, are more likely to be effective on allochromatic gems than idiochromatic gems.

PARTICOLORS

During and after gemstone growth, changes in the surroundings may result in different trace elements being available for incorporation into the gemstone. These may result in a change in color, color banding, or patches of color. When one part of a gemstone is a different color to another part, it is called particolored. Watermelon tourmaline, with its pink interior and green outer rim, is an excellent example of a particolored gemstone; others may have more than two different colors.

PLEOCHROIC (DICHROIC AND TRICHROIC) AND THE DICHROSCOPE

A gemstone may also appear different colors or shades of color when viewed from different directions (pleochroic). This is as a result of the way light travels through the crystal structure of the gemstone. When a gem shows two different colors or shades of color, such as in some rubies, the gem is described as dichroic. Trichroic gemstones such as iolite (cordierite) and tourmaline show three colors or shades of color.

To see the different colors you have to view the gemstone from one direction, remember the color, and then turn the stone and compare the image with the color when viewed from the new direction. Slight changes in color can be difficult to recognize. The dichroscope, a handheld gemological instrument with a polaroid filter and similar in size to a hand lens (loupe), can be used to view two colors of a pleochroic gem at the same time enabling comparison to be made more easily.

Peridot

Blue sapphire

Pink sapphire

Watermelon tourmaline

▲ *This view, through a dichroscope, clearly shows the pleochroism of this crystal fragment.*

The manner in which the atoms of a gemstone are arranged and the strength of the bonds between them directly affects the physical properties of gemstones, their durability (hardness, toughness, and stability), the way they break or cleave and their relative density (specific gravity).

The ability of a gemstone to withstand general day-to-day wear and tear sufficiently to be mounted in a piece of jewelry and to keep its polish without becoming unduly scratched, cracked or worn is important and will affect its value. Generally speaking, the harder the gemstone the better it can take and retain a good polish.

HARDNESS

In gemology, hardness is a measure of how easily a surface can withstand abrasion caused by wear and tear including scratching. If jewelry items containing a range of different gemstones are stored together, the harder gems such as ruby and sapphire will scratch the less hard pieces, for example emerald, amethyst, or opal.

Because hardness is related to the crystal structure and the strength and direction of the atomic bonds, the gemstone may be scratched more easily in one direction than another. This differential hardness is particularly important for the lapidary when assessing how best to cut and polish a gemstone and in choosing a suitable mount or setting for the piece.

THE MOHS SCALE OF HARDNESS

The relative degree of "scratchability" or susceptibility to scratching can be assessed using the Mohs scale of hardness. In 1822 Friedrich Mohs, a German mineralogist, took 10 minerals that he was able to obtain easily and tested them against each other. Mohs arranged the minerals in order, from the softest (talc) that could be scratched by all the others, to the hardest which could not be scratched (diamond). He then assigned them numbers from the softest (1 = talc) to the hardest (10 = diamond).

▶ *Table showing the hardness of nine materials on the Mohs scale and the Vickers scale. The Vickers hardness test or the 136° diamond-pyramid hardness test is a micro-indentation method. The indenter produces a square indentation, the diagonals of which are measured. The diamond-pyramid hardness is calculated by dividing the applied load by the surface area of the indentation.*

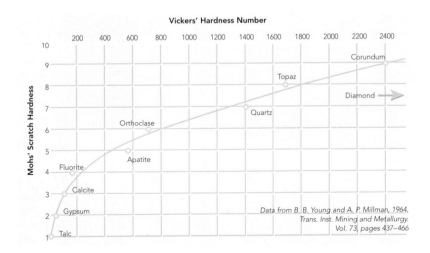

Data from B. B. Young and A. P. Millman, 1964,
Trans. Inst. Mining and Metallurgy.
Vol. 73, pages 437–466

The test is a comparative one, and the scale is not linear and does not increase by equal increments; for example, the difference in hardness between 1 (talc) and 9 (the corundum group, which includes ruby and sapphire) is less than that between 9 and 10 (diamond).

Other hardness scales include: (a) the Brinell scale, which measures how much of a dent can be made when a steel ball is pushed into the surface (this cannot be used on particularly brittle, thin or fragile pieces, as they would break); (b) the Knoop scale, which uses a diamond to make a measurable dent; and (c) the Vickers scale (see graph), which also uses a diamond to make a dent.

The Mohs test for hardness is quicker, cheaper and easier than the other methods. An estimate of hardness can be given by looking at the general wear and tear of the crystal faces (on an uncut "rough" specimen) or flat surfaces (facets) and facet edges on a cut and polished gemstone. A set of hardness pencils (each with a point made of one of the 10 mineral specimens) can be used to test hardness and is particularly useful on carved pieces, crystal fragments and pebbles.

A word of warning – hardness testing is destructive as it causes damage to the gemstone, and therefore should only be used in exceptional circumstances. Before starting to make a scratch, find an inconspicuous place and, if necessary, make a very small scratch while viewing the gemstone through a microscope or hand lens (x10 magnification).

STREAK

When a softer mineral is rubbed or scratched across a harder surface, a fine layer of the softer mineral may be removed and deposited as a fine dust or a colored mark referred to as a "streak." For example, the writing left on paper when a graphite pencil is used.

Some minerals leave a characteristic streak, for example the dark blood-red streak of hematite and the gray streak of galena. Testing for streak is destructive and is seldom undertaken by gemologists, particularly as it is of little use with small, cut stones. However, it may be useful with larger uncut specimens, rock fragments and decorative pieces, as well as to check for dying or color impregnation. For example, malachite has a green streak, but another rock type dyed green to imitate malachite may have a white streak and only a thin green dyed layer.

CLEAVAGE

The internal arrangement of atoms within a crystalline material (the crystal structure) has a regular three-dimensional pattern. The atoms may be bonded in a number of different ways, with bonds of different strengths. As a result, the crystal may have one or more planes or directions of relative weakness along which it will break more easily. These are called "directions of cleavage." When a crystalline material breaks along a cleavage plane, it will leave a flat, or nearly flat, cleavage surface. Cleavage can only occur in crystalline materials and may be defined as perfect, good, fair, or poor.

A calcite crystal has perfect cleavage

Rutile has a distinct, prismatic cleavage

An emerald crystal has poor, basal cleavage

33

Diamond is the hardest gemstone, but with a well-aimed hit, it is possible to break a diamond in two with a clean break. The clean break occurs along one of its three perfect cleavage planes.

This ability to cleave is taken into account during the fashioning of a diamond, and may be the first step in the process of cutting. The ease with which a gemstone cleaves and the number of cleavage directions it possesses are also important with regard to its durability and its identification.

FRACTURE

Fracture is a random, non-directional break that can be caused by impact, stress, pressure, or a rapid change in temperature. Fracture is not related to planes of weakness in the crystal structure and can therefore occur in both crystalline and amorphous (non-crystalline) materials.

The fracture surface may be described, for example, as uneven, irregular, conchoidal (shell-like), or hackly (uneven and jagged).

Obsidian has a conchoidal fracture

Labradorite has an uneven fracture

Silver has a hackly fracture

Kunzite has a splintered fracture

TOUGHNESS

Toughness in a gemstone is a measure of how well it can resist a fracture developing, i.e. how far a crack will propagate. Gemstones that are polycrystalline (made of more than one crystal/grain), such as jadeite, nephrite, and agate (polycrystalline quartz variety), are generally tough and difficult to break. The interlocking crystal grains are oriented in different directions, thereby slowing or stopping the propagation of the fracture.

Zircon is brittle, it is easily chipped and the corners break off as it cleaves easily. A great deal of care and attention is therefore required at all stages of fashioning by the lapidary, and later in terms of storage by the retailer and ultimately the owner.

STABILITY

In addition to physical stress due to an impact, light, heat, or some other event, chemicals may alter the structural and surface conditions, particularly of the organic gemstones such as pearl and amber. The chemicals in some perfumes, for example, will act to wear away the outer surface covering of pearls, removing their glimmering luster. For a gem to be stable, it must be able to resist both chemical and physical alteration.

RELATIVE DENSITY OR SPECIFIC GRAVITY

If you hold two specimens of the same size but of different gem species in your hand, you may notice that one feels heavier than the other. The larger the gemstones, the easier it will be for you to recognize the difference in weight. Also, if you take two gemstones of different gem species that have been cut to weigh the same, one will be larger.

The difference in weight is due to differences in chemical composition and crystal structure, which affects the way in which the atoms are arranged within the gemstone. In some gemstones, the atoms are more densely packed than in others and will have a higher relative density.

The relative density (specific gravity) of a gemstone is found by comparing the weight of the gemstone with the weight of an equivalent volume of water. It is useful in the identification of unmounted gems (those that have not yet been put in a ring, brooch, or other piece of jewelry), whether they have been cut or polished, fashioned as beads or sculptures, or simply left in their natural state (rough).

The hydrostatic weighing method of measuring the specific gravity of a gemstone is based on Archimedes' principle. The principle explains the apparent loss of weight of an object when it is submerged in water as the loss of weight is equal to the weight of the water that is displaced. The weight of the equal volume of water displaced is equal to the weight that the gemstone would be if it were made of water, so the two can be compared.

The procedure is quite straightforward: simply weigh the gemstone in air and then in water. The specific gravity (SG) of the gemstone is the weight of the gemstone weighed in air (a), relative to the difference in its weight in air compared with its weight in water (weight in air − weight in water, a − w), or:

$$SG = \frac{a}{a - w}$$

A spring balance is often used for larger specimens including carvings. More sensitive balances are needed for smaller stones.

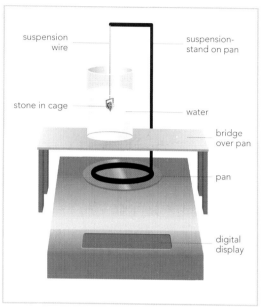

Single-pan balance used to determine a gem's relative density (specific gravity). First, weigh the gem on the balance pan and record the result (a). Then take a beaker two-thirds full of distilled water and place it on the bridge. Totally immerse the stone cage in the water and reset the balance to zero. Now insert the stone into the cage without spilling any water, immerse the cage again and weigh (w). The relative density (SG) is (a) divided by (a − w).

Gem materials have been fashioned and polished for millions of years. Where they have been left in their rough state, it may be because they look just fine as they are, and their natural beauty and desirability is worth retaining. For others, there may be a belief that the strength, soul or energy of the gemstone may be released or in someway affected detrimentally by intervention.

WORKING WITH GEMSTONES

The vast majority of gem materials, however, can be improved by the creative and practical skills of the artist and scientist. As techniques and skills have developed, the range of possibilities for the sculptor, carver, or the lapidary (the person who cuts and polishes gemstones) and others has grown, but the aim has remained the same – to work with natural materials to bring out their best qualities.

The transformation from the rough gem (such as a pebble, crystal, or crystal fragment) to the finished product (a valuable, sparkling gemstone) depends upon the expertise of the lapidary. The style of cutting or fashioning chosen will depend primarily on the optical and physical properties of the gem and the shape and condition of the rough stone.

If you are just starting out as a jewelry designer or lapidary, or would like to have a go at making your own jewelry, take time to look at examples of both rough material and material that has been cut and fashioned in a range of styles and designs. The use of the gemstones will depend partly on your creative abilities and partly on the inherent qualities of the gemstones that you choose. If you are a collector of rough and/or polished material, take time to visit gem and mineral fairs, art and craft shops and museums, and look carefully at the colors, shapes and sizes of crystals, their localities and their appearance, including any surface features (for example, striations), or internal features (for example, inclusions) of the crystal or the quality of the cut.

▼ Budding jewelry designers may start by making bracelets and necklaces using beads.

BEADS, CARVINGS, AND CABOCHONS

Gems that are heavily included or flawed, colored or patterned, opaque or translucent are generally fashioned as beads and carvings or cut as a cabochon. A cabochon consists of a domed polished top with a flat, unpolished back and has an oval or round outline. It is the simplest and oldest style of cut. It is also used to show the optical effects of iridescence, sheen, cats'-eyes (or chatoyancy), and star stones (or asterism).

Beads are generally a good place to start for the budding jewelry maker as they can be purchased fairly cheaply from arts and craft shops and gem and jewelry fairs. Beads can also be reused or recycled from unwanted or broken necklaces, etc. Possibly an even cheaper option is to make use of tumbled and polished pebbles or crystals and other rough pieces of material. Depending upon where you live or your vacation destinations, you may also be able to collect your own material (including pebbles, crystals, and shells).

FACETED GEMSTONES

In order to facet material, you will need your own lapidary equipment or access to equipment. A beginner would be wise to find a local lapidary course or join a lapidary club before investing heavily in stock or equipment. Joining a course or club will also give you the opportunity to meet others with the same interest, share lapidary experiences (both good and bad), and maybe even swap material.

Most transparent gemstones are faceted. The gems are cut and polished such that they have a number of flat polished surfaces (facets). The number of facets and the angles between them are worked out mathematically, so that the facets act as mirrors that allow the maximum amount of light to enter the stone and also be reflected back out of the front of the stone to the viewer. A well-cut gemstone will be cut to show the body color, brilliance, fire, and sparkle to its best effect. However, a compromise usually has to be found to reach a good balance between these four attributes, which will give the final appearance – referred to as the "make" of the stone. It may also be necessary to retain as much weight as possible, as the weight of the gem directly influences its value.

▲ *Jewelry making is a rewarding hobby which may be developed into a successful business.*

LAPIDARY AND JEWELRY MAKERS

When choosing the best gem material for a piece of jewelry, the lapidary and jewelry maker will be thinking of the finished product and how well fitted the material is to their specifications, or taking inspiration from the material and working out how to make the best possible piece from it. They will be assessing color, clarity, and shape as well as the durability of the gemstone, as this will affect how it can be worked, cut and set or mounted, and how it should be cared for to avoid it becoming scratched, cracked, or otherwise damaged.

The skill of the lapidary, jewelry maker or jewelry designer lies in their ability to recognize the qualities of the gem materials and work with them to their best abilities to produce a piece that is both admired and sought after. They will need to find the best compromise between what is possible and what is practical, whether they are sawing, grinding, sanding, lapping, polishing, drilling or tumbling, to produce beads and spheres, cabochons, faceted gemstones, inlays or mosaics, cameos or intaglios, carvings or sculptures.

As a rough guide to the care needed to work gem material and the difficulty in working with a material:

- The tougher (more durable) the material, the less likely it is to fracture, but the more brittle, the more likely it is to chip, crack or shatter.
- The better the cleavage, the more likely it is to break cleanly along a cleavage surface.
- The softer the material, the easier it is to scratch or carve, but the harder it is, generally the better the polish it will take.

DURABILITY AND HARDNESS

Hardness is a measure of "scratchability"; toughness is a measure of how resistant a gem material is to attack by physical and chemical processes. A gemstone such as spinel with a hardness of 8 (on Mohs' scale of hardness) is relatively hard, but because it is also brittle and easily broken extra care must be taken when handling or working with it. In contrast, a gem material like jadeite with a hardness of less than 7 is exceptionally durable. Corundum has a hardness of 9, so hard that the prongs or setting may be damaged before the gemstone. When cleaning or working with corundum jewelry, check that the gemstone is secure in its setting. Tourmaline is both brittle and fragile, and care must be taken to provide a setting that protects the gemstone. Rubover settings and claws will help protect the edges of fragile gemstones.

Quartz and chalcedony have a hardness of 7, are quite durable and relatively easy to work with. They are available in large sizes, abundant and fairly inexpensive, and can be engraved, carved or faceted. Lapis lazuli (hardness 5–6) and malachite (hardness 4) are versatile rock materials and can be worked in a number of ways, including sawing, carving, and faceting.

▶ *A jewelry maker sorting through gemstones, looking for those of a similar shape and size.*

FRACTURE AND CLEAVAGE

Topaz has perfect cleavage in three directions and particular care should be taken when cutting, polishing or setting to avoid breakage. The emerald cut or octagonal cut is mainly used for emeralds; the corners are removed during faceting so that they cannot be knocked off accidentally while being set or later when the finished product is being worn. Rhodochrosite is also liable to break along planes of cleavage.

Agate is banded may break parallel to the banding, so care should be taken. The layering in feldspars may also weaken the material, so care should be taken. Chalcedony and agate, are porous and will easily take a dye or stain, to change or enhance their color, but may just as easily become stained unintentionally by chemicals or oils, etc., so care must be taken when handling and working with them. Care should be taken with other porous gemstones, such as turquoise and opal.

COLOR AND SHAPE

Irregular-shaped tumbled gemstones, pebbles, beads and crystal fragments can inspire a jewelry designer. A well-proportioned and well-cut gemstone is easier to work with than an irregular or uneven faceted gemstone. Where a gemstone is cut to give an unbalanced gemstone, either too shallow or too deep, care should be taken to ensure that the gemstone is held securely, and that the prongs and setting do not accentuate the lack of balance in the gemstone. A gemstone may have been cut particularly deep in order to improve a pale or insipid-colored gemstone, to add weight to a gemstone, or possibly to avoid using a fractured or heavily included part of the original crystal or pebble. A dark-colored gemstone may have been given a shallow cut, for example, in order to brighten the color. These gemstones may need to be recut.

Gemstones occur in a range of colors and finding matching gem material, for example, for a pair of earrings is easier for some gem materials than others. Tourmaline has a very wide range of colors and matching can be particularly difficult. It may be easier to use rough material from which to fashion two matching pieces rather than try and match already faceted gemstones from different crystals.

CLARITY

The position, type, and number of inclusions in a gemstone affect the durability of the gemstone and will affect how it is fashioned and set. Where inclusions are close to, or reach, the surface they may cause an uneven surface or weaken the gemstone. Inclusions such as the "lily pad" inclusions in peridot may cause internal stress

▲ *Topaz crystal (left) and a faceted topaz (right).*

and so weaken the gemstone. Care should be taken when working with included gemstones such as emerald and a setting chosen that offers adequate protection.

CAT'S-EYE AND STAR STONES

Care should be taken to orient gemstones that show optical phenomena, for example star stones, cat's-eyes, or materials showing iridescence. A star stone or cat's-eye should be cut as a cabochon and set to show a well-centered cat's-eye or star. Check the base as well as the dome of a cabochon as a cabochon of chrysoberyl, garnet or chalcedony, etc., may have a rounded or uneven base which makes it more difficult to set.

PLEOCHROISM

The orientation of a pleochroic gemstone, one showing more than one color or shade of color when viewed from a different direction, is important and care should be taken to examine the gemstone from all angles in order to choose the best way to display or set a gemstone in a piece of jewelry to show its best color. Tourmaline, for example, is strongly pleochroic and should be examined carefully.

MAKING A CHOICE

The lapidary or jewelry maker will take these attributes into account when choosing a gem material and working with it, and should also be aware of the effect of water, acids and other chemicals, strong heat or light, etc., on gem materials so that care can be taken in handling and working with them (see *Caring for Gemstones and Jewelry*, pages 68–9).

Jewelers also need to understand the strengths and weaknesses of various gem materials in order that they can be confident of the identity of the material they are trading and also so that they can advise the customer on the best care and cleaning methods (see pages 68–9), add interest by providing details on where the gems may have formed and most importantly disclose any treatments for example oiling or heating, which may affect the value, use or durability of the article.

KEY

Toughness (durability)

*   *poor or fair*

**   *good*

***   *very good*

Cleavage

X   *no cleavage*

*   *cleavage, indistinct*

**   *cleavage good, take care*

*** *perfect cleavage in one or more direction, take particular care*

	TOUGHNESS	CLEAVAGE	NOTES
AMBER	*	X	Used as tumbled stones, beads or for carving.
AMETHYST	*	* or **	Faceted or used as tumbled stones or beads, cut as cabochons or used for carving.
AQUAMARINE		*	Faceted or used as tumbled stones or beads.
CHALCEDONY (INCLUDES AGATE)	***	X	Used mainly for tumbled stones, polished slabs, beads, pendants, and carvings.
CHRYSOBERYL	** to ***	X or *	Faceted or cut as cabochons.
CITRINE	**	* or **	Faceted or used for beads and carvings.
CORAL	* to **	X	Used mainly for beads, pendants, earrings, necklaces, cameos, and carvings. Scratches easily.

	TOUGHNESS	CLEAVAGE	NOTES
DIAMOND		***	Diamond cutting equipment needed. Faceted.
EMERALD	* to **	*	Emerald (and beryl) toughness depends on clarity. Faceted or used as tumbled stones or beads.
FLUORITE	*	***	Cleavage perfect and lack of toughness, used mainly for beads, pendants, earrings, necklaces, and carvings.
GARNET	**	X	No cleavage, but may have indistinct parting. Faceted or used as tumbled stones, beads or cut as cabochons.
HEMATITE	*	X	High density makes it heavy for jewelry, used mainly for beads, pendants, and carvings.
IVORY	*	X	Used mainly for beads and carvings. Scratches easily.
JADEITE AND NEPHRITE JADE	***	X	Jade is particularly tough. Nephrite jade is even tougher than jadeite jade. Mainly used for beads, pendants necklaces and carvings.
JET	* to **	X	Used mainly for beads, pendants, earrings, necklaces, and carvings. Scratches easily.
LAPIS LAZULI	*	X	Used mainly for beads, pendants, earrings, necklaces, and carvings.
MALACHITE	*	X	Used mainly for beads, pendants, earrings, necklaces, and carvings. Scratches easily.
MOONSTONE AND AMAZONITE FELDSPARS	*	***	Cleavage perfect or easy in two directions, used mainly as beads and cabochons.
OPAL	*	X	Irregular shaped pieces used in jewelry, cut as cabochons or carved.
PEARL	*	X	Used for necklaces and earrings.
PERIDOT	* to **	*	Faceted or used as tumbled stones and beads.
RHODOCHROSITE	*	X	Used mainly for beads, pendants, earrings, necklaces, and carvings.
RHODONITE	*	X	Harder than rhodochrosite, used mainly for beads, pendants, earrings, necklaces, and carvings.
RUBY	***	X	Faceted or cut as cabochon.
SAPPHIRE	***	X	Faceted, cut as cabochon or carved.
SHELL	*	X	Used mainly for beads, pendants, earrings, necklaces, cameos, and carvings. Scratches easily.
SPINEL	**	*	Faceted or cut as cabochon.
TANZANITE	*	***	Faceted. Cleavage perfect in one direction.
TOPAZ	*	***	Faceted. Cleavage perfect in one direction.
TOURMALINE	*	X	Faceted or used for tumbled stones, beads, pendants, earrings, necklaces, and carvings.
TURQUOISE	* to **	X	Used mainly for beads, pendants, earrings, necklaces, and carvings.
ZIRCON	* to **	*	Faceted.

Art Deco diamond brooch (c. 1930), set with old European-cut diamonds

In the gemstone trade, diamonds are usually treated separately from other colored gemstones. They are the hardest gemstone and are generally cut and fashioned by diamond cutters, who will specialize in their manufacture, or even just one stage in the diamond cutting and polishing process.

The earliest diamond cuts probably entailed no more than removing the top of an octahedron crystal, known as the point cut. Another early cut is the table cut, where the point of the octahedron is flattened to a square facet called the table. As cutting techniques developed, further cuts were made and the polishing of facets was introduced. The rose cut, with a flat back and domed and faceted front made of triangular facets, was introduced by the 14th century.

The most common cut for a diamond is the brilliant cut. The first type of brilliant was the single cut (or half cut) with the main table facet surrounded by eight smaller facets on the crown (the upper part of the stone) and eight on the pavilion (the lower part of the stone). During the 17th century, this evolved to the double cut (or Mazarin cut), with 16 crown and pavilion facets and a culet (the small face formed by polishing the point of the pavilion to a flat facet). The number of facets continued to increase. By the middle of the 17th century, the triple cut (or Peruzzi cut), and variations including the Brazilian cut and Lisbon cut, had 58 facets (including the culet) – the same number as in the modern-day brilliant, which was introduced in the 1820s along with mechanical bruting (see page 45).

BRILLIANT CUT

The proportions of the modern brilliant cut were developed by a number of cutters, among whom the most influential was Henry Morse, who in 1860 opened the first diamond-cutting factory in the United States at Boston, Massachusetts. Other designs have since been introduced, including the "round brilliant" of Marcel Tolkowsky, whose book *Diamond Design* (1919) systematically analysed the optics of a diamond to show the best fire and brilliance. The brilliant cut is also used for many other gemstones. Variations continue to be developed, usually in accordance with the designs for standard proportions of Eppler, Scandinavian Diamond Nomenclature (SCAN DN Cut), and the International Diamond Council (IDC).

Table cut

Brilliant cut

Round brilliant-cut diamond gem

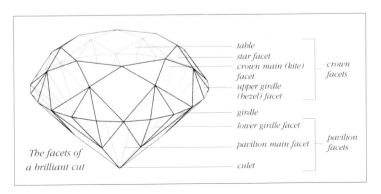

table
star facet
crown main (kite) facet
upper girdle (bezel) facet
crown facets

girdle
lower girdle facet
pavilion main facet
pavilion facets

culet

The facets of a brilliant cut

FANCY CUTS

Fancy cuts include variations on the round brilliant cut, such as the oval, pear or drop-shaped (pendeloque), and boat-shaped (navette or marquise) brilliants in addition to triangular, hexagonal, pentagonal, kite-shaped, etc. With advances in computer technology and lasers, new shapes and styles are entering the market, including computer generated designs and gems engraved by laser.

STEP CUTS

Step-cut (also known as trap-cut) gems are generally cut to show their color rather than fire or brilliance. In gemstones that show pleochroism, the gem is oriented and cut to show the best color when viewed through the table facet. Sapphires and other gems that may show a color banding are oriented so that the banding is parallel to the steps of the cut and cannot be seen clearly through the table. If there is an area of a stronger color, it is best placed toward the base of the gemstone, deep within the pavilion, so that the color appears to flood the whole stone when viewed through the table facet.

The step cut has a square or rectangular table facet and girdle, surrounded by a series of parallel four-sided facets. The number of facets is determined by the size of the gem. In brittle gems, such as emerald, the corners may be easily damaged, so they are usually removed during cutting. The resulting cut is called the emerald cut or modified step cut. As more of the corner is removed, the gem nears the outline of a regular octagon and may be called an octagonal cut.

Pendeloque-cut pyrope garnet

Step-cut goshenite gem

Baguette-cut tourmaline gem

Cushion-cut cuprite gem

CUSHION CUT AND MIXED CUTS

In order to get the best "make" and maintain a good weight, the cushion cut is often used. The cushion cut is a variation of the brilliant cut, with rounded cushion-like appearance. Mixed cuts optimize both the color and the fire of a gemstone by using different cuts on the crown and pavilion facets. The most common mixed cut is a brilliant-cut crown with a step-cut pavilion.

STAGES IN FACETING A BRILLIANT-CUT DIAMOND

The main diamond-cutting centers are Antwerp, Mumbai (formerly Bombay), New York City, and Tel Aviv (Israel). Although much of the process can now be carried out

Marquise

Pendeloque

Fancy

Step

Baguette

Cushion

Mixed

▲ *Brilliant-cut treated yellow diamond ring, weight 10.12 carats, with a two-tiered marquise-cut diamonds.*

with the aid of computers, there is still a need for the "eye" of the expert to check the progress and assess the accuracy, make, and overall look of the gemstone. A different expert may carry out each step of the process.

STAGE 1   The designer will decide how best to use the shape of the rough stone, taking into account any flaws or inclusions, the weight, and the overall quality of the stone. The designer may be looking to produce the largest stone possible, or several smaller stones with a larger combined weight. There will usually be a compromise between the number and size of the stones that could be produced and the final yield. The value and quality of the polished diamond will be judged on its color, cut, clarity, and carat (weight) – the 4Cs.

STAGE 2   Once the decision has been made how best to use the rough stone, it is marked for the first cut (saw) or cleavage (a break along a plane of weakness related to the atomic structure of the diamond). Nowadays, diamond sawing can be carried out by in a factory by computer-controlled lasers.

STAGE 3   The first step in faceting an octahedral diamond crystal may be to saw off the top of the crystal using a rotating disc coated with a coarse diamond abrasive. A diamond may also be cleaved to divide it in two. In order to cleave a diamond, the diamond is first cemented to a stick (dop), a small groove (kerf) is made along the mark into which is placed a steel blade. The blade is then given a sharp hit with an iron bar to cleave the diamond in two.

STAGE 4   The diamond is given a rounded outline by holding it against a revolving cast-iron wheel, known as a "scaife," coated with a

▶ *Gem polisher at work in Thailand.*

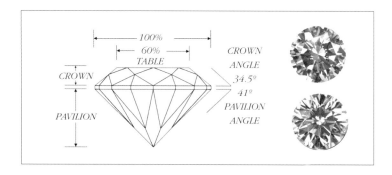

CROWN
ANGLE
34.5º
41º
PAVILION
ANGLE

100%
60%
TABLE
CROWN
PAVILION

◄ At far left, the ideal
proportions of a brilliant-
cut diamond. At left, two
brilliant-cut diamonds:
the larger table (top) has
less reflection, while light
is reflected upward in the
smaller table (bottom).

diamond powder paste. This process, called "bruting," gives the diamond the appearance of frosted glass. The girdle, the widest part of
the stone, which will separate the top (crown) facets from the lower
(pavilion) facets, can be seen at the end of this stage.

STAGE 5   The facets are then made using the scaife. The flat table facet
is ground first and then the bezel facets. The facets are ground and
polished in a particular order; first the pavilion main facets, the culet,
star facets, upper girdle facets and lower girdle facets.

STAGE 6   A "brillianteer" then adds the remaining 24 crown and 16
pavilion facets, bringing the total to 58 facets including the culet. A
finer-grained diamond abrasive powder is used for each progressive
stage of polishing. The diamond is then ready to be mounted in a ring
or other piece of jewelry.

CHOOSING THE MOST SUITABLE CUT

When choosing the best cut for a gemstone that is being fashioned
to be set in a piece of jewelry and sold by a high street retail outlet
or jewelry shop, a lapidary will look for a compromise between
appearance and size (carat weight) as the value of the piece will
depend partly on size. For colored gemstones, the color is the
most important attribute, and a good colored gemstone will have a
premium. For designer jewelry, the gemstone may take second
place, with the value added by association with the fame of the
designer and the appearance of the setting. Where gemstones are
used, they should be well cut.

An appraisal of the cut of the gemstone is referred to in the
jewelry trade as the "make" or "cut." A gemstone with a good make
or cut with symmetrical, facet edges should be sharp and given a
good polish and the angles of the facets should be fashioned so that
as much light as possible is reflected back through the front of the
gemstone, giving the gemstone the best possible sparkle and fire.
The gemstone should look well-proportioned, neither too deep or
too shallow, or top or bottom heavy. The "shape" of a gemstone is
the term used to describe the outline of the faceted gemstone: for
example, round (as in round brilliant cut), oval, square, rectangle, or
fancy shapes such as heart and kite.

BUYING A DIAMOND

The value of a diamond is said to be in its ability to evoke emotions of love and romance. Diamonds are most often given as a token of love, so selecting a diamond is important to both the customer as well as the recipient. Most diamonds are sold at Christmas and Valentine's Day, a day associated with romance. Many engagement rings contain a single diamond (solitaire) or a colored gemstone and one or more diamonds. The most popular cut is the round brilliant, a traditional cut with perfect symmetry and a high degree of sparkle. Customers will pay between 10–15% more for a well-made round brilliant cut. Other cuts such as princess, emerald, baguette, marquis, and pave cuts are a more personal choice.

Buying a diamond may be an impulse buy, but if possible take time to learn more about the diamond industry, get to know local jewelers and spend time looking at diamonds, reading about them and talking about them.

When choosing a diamond for yourself or as a gift for a loved one, remember that the cost of the diamond is decided following an assessment of the color, cut, carat (size), and clarity of the stone. There will also be other costs to take into account too, such as the value added by the setting, the fame attached to a jewelry designer, and also overheads needed to cover a proportion of the costs incurred by the shop, such as rent, heating, lighting, and staff costs. The mark-up between the wholesaler and the retailer may be as much as 300%. The mark-up from the value of the rough to the finished

▼ A row of rough diamonds arranged in order of clarity (top) and color. When being checked for clarity, diamonds are given grades ranging from flawless to heavily included. When checked for color they are given grades ranging from "D" (colorless) to "Z" in normal range of yellow, gray, and brown. Outside this color range diamonds are considered "fancies" if they have a strong color. The Diamond Trading Company sorts diamonds into around 12,000 categories of clarity, color, carat weight, and cut, known as the 4Cs.

product is probably somewhere in the region of 400% or 500%. The reason why gems are typically priced so high to the ultimate consumer is because of the number of "middle men" each taking a cut in order to run a profitable business.

Diamonds and colored stones follow a similar route once they reach the importer, passing from the wholesaler to the manufacturer, the wholesale jeweler, to the retail jeweler and finally to the customer. The same diamond may cost more in a well-known jeweler's shop in the center of a capital city than in a local town.

WHAT TO LOOK FOR?

You may be looking for a particular color or shape or may be just concerned with size. If possible, have a look at a number of different diamonds. A diamond that is certified as having very small inclusions (VS1 or VS2) which are difficult to see without a x10 hand lens will cost less than a diamond which has inclusions that are difficult to see even for an experience diamond grader using a x10 lens (VVS1 or VVS2), and far less than a flawless diamond which shows no inclusions under a x10 lens.

Most people are unlikely to use a x10 when wearing a piece of jewelry or noticing a piece of jewelry worn by a friend or colleague, and even quite a large inclusion (SI1 or SI2) is unlikely to be noticed across the candle-lit table. Unless the diamond contains obvious inclusions which can sometimes be seen easily with the naked eye (I1, I2 or I3) it is unlikely that the inclusions will be noticed even by the most curious admirer.

Some jewelers talk about the 20% rule. If you are looking at two diamonds of the same shape and quality (color and clarity), to look larger, one must be a minimum of 20% more in carat weight. This is known as the 20% rule and is important, as there is little point in buying a slightly heavier stone, which will cost more, if it looks no larger. Unless, of course, you just want to be able to quote the carat weight of the diamond you buy, in which case the heavier the better.

ARE YOU ONE OF THESE BUYERS?

1. Size is most important – the bigger the better
Color and clarity don't matter to you.
Go for the yellowish diamond with inclusions – as long as the yellow isn't too obvious and the inclusions don't detract from the beauty of the stone too much.
How about 1 carat and larger, I2 clarity and L–M color.

2. Size is important – but it has to look good too
Color and clarity are important, but not the main concern.
Go for the slightly yellowish diamond with inclusions that are so small they are hardly noticeable.
How about 0.75 carat and larger, I1 clarity and K color.

3. Size and quality are equally important
The diamond doesn't have to be perfect, but it should look really good.
How about 0.50 carat and larger, SI1 clarity and I–J color.

4. Quality is more important than size
The diamond must not only be eye-clean, it has to be reasonably clean when examined under a jeweler's loupe (×10 magnification). It must be white and sparkle.
How about 0.50 carat or larger, VS1 clarity and G color.

5. Profit is most important, so the right shape is essential
For an investor who aims to store the diamond and resell it at a later date for a profit, a round brilliant-cut stone is best.
How about 1 carat and larger, VVS1 to FL clarity and D, E, F color.

The diamond should come with a diamond-grading laboratory report from a reputable laboratory.

▲ *Inclusions are not uncommon in diamonds. This image shows the diamond magnified x10.*

So, think about what you are happy to buy – do you want to know that the diamond you give is flawless, the biggest or the best, even though the receiver probably will not notice the difference? Do you want to know the value, the clarity and color grading, and the carat weight of the diamond that you receive, or are you content to have a diamond that is given by someone special and looks just fine?

Success in the diamond industry is based very much on reputation and personal contacts. Many jewelers and gem dealers can trace their trade through several generations. Nowadays it is even more important for a reputable local jeweler to be aware of what he/she is selling and to offer a fair price in order to compete with internet sales, shopping channels and other large ventures. Choose a jeweler with whom you feel confident, meet regularly, and ask questions.

CERTIFICATION

If you are buying a particularly large or expensive diamond you may wish to check that it has been certified by a gemological laboratory and that the jeweler is certain of the identification. Generally, however, a certificate is not necessary unless you are a diamond investor buying investment-grade diamonds and intend to sell them on at a later date, or you are a diamond investor buying fancy-colored diamonds and need certification that the color is natural, or you are concerned that the jeweler might exchange your diamond for a different stone (synthetic or imitation) or a diamond of lesser value, for example, when resetting a stone, cleaning a stone or modifying a piece of jewelry. In the final scenario, if you are worried, ask the jeweler to plot internal characteristics (inclusions, flaws, and so on) of your diamond, take its measurements, and sign these on the job sheet before the work is done. Have these checked again after the work is done. If you are really worried, go elsewhere.

For a diamond report to be valid, it must be recent (within six months) and the diamond must not have been worn since certification. If you are having diamonds certified to guarantee current quality before selling them, do not allow them to be worn as this could invalidate the report.

There are many synthetics, simulants and imitations on the market, and without up-to-date training and knowledge it is difficult for many jewelers to keep aware of the modern technologies. Ask whether the jeweler is a member of a trade organization and whether their staff receive regular industry updates and training.

SPOTTING A FAKE

What can you look for? Without having seen a number of natural diamonds and diamond simulants, it is difficult to spot the difference between a diamond, CZ (cubic zirconia), moissanite, or another colorless stone designed to imitate a diamond. It is even more difficult when the stone is set in a piece of jewelry. Comparison of a CZ and a diamond will show that the fire (sparkles of different colors) in a CZ are brighter and somehow "too much," compared with the subtle but wonderful rainbow shown by a diamond when it is moved with a light shining on it. Glass (paste) looks dull and lifeless compared with diamond. There are a number of other tests, though most need to be carried out using gemological equipment. The "spot test" and "tilt test" are a couple of tests you might like to try at home.

The spot test is more easily done on an unset round brilliant-cut diamond, but may also be possible where the diamond is not set in a closed setting, i.e. the culet (the point on the back of the stone) can be seen clearly from the back of the piece of jewelry. The band of a ring would usually make it too difficult to get a good enough view of the back of the stone to test it. If the diamond is not set in a ring or a piece of jewelry and the cut (or make) of the stone is good, i.e. the stone is well-proportioned, symmetrical and has been cut well, no light should leak out of the back of the stone. Draw a spot on a piece of white paper and place the diamond table facet down on top of the spot. The black spot will not be visible through the diamond but will be visible through a brilliant round-cut CZ, paste, quartz (rock crystal), spinel, white (colorless) sapphire, moissanite, or other diamond simulant.

The tilt test can be carried out on a diamond that is set in a ring or a piece of jewelry, as long as the edge of the table facet (the large flat face on the front of the stone) can be seen clearly. Hold the diamond so that the table facet is horizontal. Look along (not at) the surface of the table facet and then gradually tilt the stone toward you. This test is similar to the spot test, and if, as you tilt the gemstone, you see a dark, fan-shaped area through the table facet, it is unlikely to be a diamond.

If you are concerned about the source of the diamond and wish to support ethical sourcing, ask the jeweler where he/she buys the diamonds, and do they buy from a reputable source? Can the wholesaler follow the route back to the cutter/polisher, and can they confirm that the rough material was certified using the Kimberley Process Certification Scheme?

WHAT SHOULD YOU PAY?

There isn't a right amount. There have been some studies of the amount that men spend on engagement rings, with results ranging from a proportion of their monthly income in some countries to five or more times their monthly income in others, but as much depends on personal circumstance and more often than not a totally free choice to buy the perfect diamond is just not possible. Who buys diamonds? They are sold in practically every country of the world. However, as a luxury item, they are more likely to be purchased in developed countries than those where getting enough food and fresh water are the main concerns. Most of the diamonds are produced in Third World countries for sale in developed countries. The US, with only 5% of the world's population, absorbs about 50% of the global diamond market.

If you are a jewelry designer, you will be concerned not only with the 4Cs, but will be looking for other attributes, looking for how best to use the diamond or diamonds to create a piece of jewelry that demonstrates creativity, originality, wearability, and craftsmanship. The color, shape, and size serve as a source of inspiration.

▼ *A pair of engagement rings featuring diamonds set in a platinum ring.*

Many people are not aware that there are still new gemstone sources being discovered and that sources which have not produced for a while are starting again. An overview on what is new or newly available helps dealers and retailers adjust pricing and get better samples. Keeping up to date with new discoveries and knowing what new gemstones are available will help you decide which gemstones to buy and which sources to trust.

Most importantly, buy from a reputable dealer or retailer, buy what you can comfortably afford and what you like. If you are buying for investment, ensure that the gems that you are buying for your portfolio have the best color, clarity, and quality that you can afford. If possible, get advice from a gemologist before investing. Avoid buying without viewing the stones and if possible view them both in daylight and artificial light, as the intensity and hue of the color may vary depending on the light source. Some stones may also be color-change stones and show quite different colors or shades of color when exposed to sunlight or artificial light. Colored gemstones, particularly rough material (including crystals) should also be viewed from all sides as color and intensity may vary (see pleochroic gemstones, page 31).

When buying colored gemstones, the color is the most important attribute and may count for as much as 50% of the value of the stone. Clarity counts for about 25–30% of the value. Inclusions and flaws may detract from the appearance of the stone and lessen its value, but interesting, attractive or characteristic inclusions may add to the value, for example the characteristic "horsetail" inclusions found in green demantoid garnet and rutile inclusions seen in colorless quartz (rock crystal). The cut is subjective and may have little or no effect on the price. The remaining value is based on carat weight, so when buying colored gemstones, it is better to buy for beauty rather than size.

COLOR

As with diamonds, where it is important to make the distinction between color grading, which can be carried out on unmounted diamonds, and a color assessment which is carried out on diamonds set in jewelry, the color of a gemstone will be affected by the color and reflectance of the setting and the light under which it is viewed. The setting and the light should therefore be taken into account when assessing the color of any colored gemstone.

The intensity of color is important, for example the intensity of the green color of an emerald will greatly influence its value. A pale or insipid green will be less valuable; a green beryl will cost less than an emerald. The most prized emeralds are mined in Colombia, South America. Recently, new deposits of emeralds have

▼ As with all colored gemstones, the color of an emerald will affect its market value; the richer the green, the more valuable the gem. This is a 75-carat emerald, set in platinum and surrounded by 129 diamonds.

been found in Zambia and Zimbabwe in East Africa. However, they are not as accepted as the Colombian gems, and tend to sell for a 50% discount to the Colombian gems.

The color of gemstones may also be associated with the locality; for example, rubies from Burma (Myanmar) have a premium, as do Kashmir sapphires. Sapphires from Burma have a stronger blue color than those from Sri Lanka or Thailand, though heat-treatment can enhance a stone lacking in intensity or of a lesser color. The color that is associated with a particular gem may be usual or rare, for example sapphire occurs in many colors and can be treated to change or enhance its color.

Natural untreated pink-orange of padparadscha has a premium as it is unusual compared with blue sapphire. Premiums for unheated Burmese and Kashmir sapphires and unheated Burmese rubies are expected to increase as some dealers and collectors are avoiding heated fancy-colored sapphires due to the beryllium diffusion treatment (see *Gemstone Enhancement*, pages 61–3). Unheated stones are now commanding 40–50% premiums over heated goods. It is difficult to identify gemstones that have had this treatment and even professional dealers are generally advised not to buy corundum without a report from a major independent laboratory. Be especially careful of intense orange, vibrant yellow sapphires and pink-orange padparadscha.

In opal, the stones that show flashes of red are generally more valuable than those that show only blue and green, as red is more unusual. It may be, however, that you find the flashes of blue and green more attractive and would gladly spend more on a piece that you particularly like, rather than a smaller piece that is more expensive. The same could be true for any other gemstone that shows an optical effect or is special to you in some way. It may even be a pebble or a crystal of a particular color or shape that catches your eye rather than the most expensive piece of jewelry in the shop.

Most diamonds are colorless with a yellowish tinge or a hint of yellow. Colorless diamonds are more valuable than those with a yellowish tinge. Fancy-colored diamonds are assessed on the intensity of their color. Red diamonds are the rarest and therefore the most valuable, though red diamonds can now be produced from lower-value diamonds by high-pressure, high-temperature (HPHT) methods (see *Gemstone Enhancement*, pages 61–3). After red, comes green, blue, pink and yellow. Yellow to colorless diamonds can be valued between hundreds to thousands of US dollars per carat. Some fancy-colored diamonds, for example pink diamonds, may be valued in the region of tens of thousands of US dollars per carat.

In general, gemstones that are in bright colors, such as red, blue, green, and yellow, are more valuable than those that are dark

▲ *Pink and blue rough diamonds. These colors are rare in diamonds and are known as "fancies." Pinks are found in Tanzania and Australia, and blues from Cullinan in South Africa.*

colored, such as brown, gray, or black. If the gemstone is heavily included or opaque, the color may have less influence on the value. There are exceptions and some gemstones are only found in these darker colors, for example hematite (metallic gray in color) and jet (black). The low price does not necessarily mean that the gemstone is unattractive: it may be that it is more readily available than another gemstone.

CLARITY

Clarity in diamonds is one of the main criteria used in valuations, along with color, cut, and carat weight (the 4Cs), and grading for clarity follows specific guidelines and grading schemes (see page 46). The appearance and quantity of inclusions or flaws in colored gemstones is less prescriptive. Much depends on where the inclusions are and whether they are easily seen. Attractive, distinctive or characteristic inclusions can add value to a gemstone. They may also serve to show that the gemstone has a natural origin rather than being a product of a laboratory. Laboratory-produced artificial gemstones and synthetics will not contain naturally formed inclusions. They may even look "too clean to be real." The value is in the overall appearance of the gemstone rather than its clarity.

Optical phenomena such as cat's-eyes and star stones are formed as a result of inclusions. These may add value and interest to an otherwise translucent or opaque gemstone. Other optical phenomena that may add value to the gemstone include: adularescence (e.g. moonstone); labradorescence (e.g. labradorite); aventurescence (e.g. aventurine quartz); iridescence (e.g. hematite). When buying a gemstone that may show these optical phenomena, be sure to view the gemstone in both natural and artificial light and to move the light source, or the gemstone, as you look at it.

For gemstones such as ruby and sapphire, a translucent or opaque gemstone is unlikely to be faceted and will be cut *en cabochon*, or fashioned as a polished slice. These are of far less value than a faceted transparent piece that has been well cut. Unfaceted gemstones generally cost less than faceted pieces as they cost less to produce and material of a lower clarity can be used.

Cut

The value that is added or taken from the gemstone by the quality and type of cut chosen is subjective, as it will depend on how it affects the attractiveness of the gemstone. A poorly chosen cut may not show the gemstone at its best, and a poorly or badly cut gemstone may be discounted by as much as 50% in order for the dealer to sell it. A new or fashionable cut or a particularly attractive or well-cut piece may be sold at a premium.

▼ *Fine and poor examples of gem cutting and polishing: a well-cut citrine (yellow) and a poorly cut sapphire (blue).*

Carat weight

Gemstones are usually sold by weight (carat weight) rather than size. Generally the higher the carat weight, the more valuable the piece. However, with colored gemstones, as noted above, there is more to the value than just the weight. When buying gemstones, a dealer may have separated faceted gemstones of a particular type into categories, each with a per-carat value. To work out the cost of the gemstone multiply the carat weight by the per-carat cost. By comparing per-carat prices of transparent gemstones rather than total price, it is easier to compare prices accurately. Also, when comparing gemstones try to compare "like with like," that is, gemstones of the same color, size, shape, and quality. Gemstones such as amethyst and blue topaz occur in a wide range of sizes and each may have the same or similar per-carat price. Gemstones such as ruby, sapphire and emerald, which are seldom found as large gemstones, will have a higher per-carat price for the larger gemstones than for the smaller, more usual, sized pieces.

Two gemstones of equal weight can be different sizes. This is because of the difference in density; for example, a 1-carat ruby and a 1-carat emerald each weigh 1 carat, but the emerald will be larger because it has a lower density (see specific gravity, page 35).

It is also important to note whether the carat weight is that of a single gemstone or of the total within a piece of jewelry or in a purchase. Larger gemstones are more rare and therefore more valuable. Check carefully to see whether the label says "1-carat total weight" (TW) or "1 carat," the weight of the gemstone. There is also a premium for a gemstone that is cut to a full weight, for example 0.5 carat or 1.0 carat, and the lapidary will bear this in mind when cutting and polishing a gemstone.

▲ *Diana, Princess of Wales, wearing a suite of sapphire and diamond jewels presented by the Crown Prince of Saudi Arabia, and the Spencer family tiara.*

THE PRICE OF COLORED GEMSTONES

Gemstones fall broadly into three categories: those that one might buy for everyday wear and are fairly affordable; those bought for special occasions; and the very expensive gemstones that only a very small minority would ever have the opportunity to own. Values and prices vary considerably, so it is not possible to give a definitive list.

Just a few examples of the better-known gemstones are given on page 54 to show the wide range in value, from gemstones costing less than US$200 per carat to those costing tens of thousands of US dollars per carat for a well-cut gemstone of good color and quality. This table is only a rough guide to comparative values. Particular gemstones may also be valuable for reasons other than color, cut, carat, and clarity, for example those that are associated with a particular member of royalty, a film star or celebrity, a particular locality or famous mine,

COLORED GEMSTONE PRICES

Gemstone			
Alexandrite		**	***
Almandine garnet	*		
Amethyst	*		
Aquamarine	*		
Black opal		**	***
Citrine	*		
Demantoid garnet		**	***
Diamond		**	***
Emerald		**	***
Fire opal	*		
Irradiated blue topaz	*		
Irradiated blue zircon	*		
Labradorite	*		
Paraiba tourmaline		**	***
Peridot	*		
Pyrope garnet	*		
Ruby		**	***
Sapphire – natural	*	**	***
Sapphire heat-treated or synthetic	*		
Star ruby	*	**	***
Star sapphire	*	**	

▲ *Comparative gemstone values for a well-cut gemstone of good color and quality, from less than US$200 per carat (*) to tens of thousands of US dollars per carat (***).*

a historical event, or a record breaker such as the largest, the best color, the heaviest, and so on.

In order to keep up to date and follow trends in the market, gemstone traders and valuers work from lists that are updated daily or weekly. A gemstone's value may vary depending on any number of reasons including fashion, availability of stock, the discovery of new sources and the closure of old sources, changes in import and export duties and taxes, the introduction of new treatments, and conflict in areas of gemstone mining or production.

BUYING ABROAD

Although it is usually best to buy from a reputable jeweler near to home, so you can return any purchase, many people like to buy while traveling abroad. Beware and buy only what you can comfortably afford. It can be fun to buy while traveling in your own country or abroad, and many people return with beautiful mementoes of a special visit or holiday. However, others may have been pressurized into buying gemstones by traders well experienced in the "art of the hard sell," or buy out of desperation, for example just as a tour bus is about to leave and the pressure is on to make a quick decision. They may regret such a hasty purchase.

Relying on buying gems as a means of financing a trip is inadvisable, for example students spending more than they can comfortably afford in the hope of selling the gemstones on their return may find that the experience does not turn out to be a happy one. It is extremely unlikely that gemstones bought in this way will cover the finances of the trip, and without an understanding of import/export laws and taxes, the costs entailed may even exceed the original budget.

As long as you buy gemstones that you are pleased with and can afford, then any bad news, such as the stones are not worth as much as you paid for them, or that they are fakes, simulants or synthetics, should not leave you too much out of pocket, merely a little wiser.

Most countries have reputable high-street jewelers. However, if you wish to be a little more daring and bring something back that you might not normally find in your own country or something that is typical of the country, such as a souvenir of a particular place, then the following section on traveling abroad may help you decide what you might like to look out for. You may even have time to look for and collect your own specimens. Even if you find what you think you are looking for, beware, as the new initiate is always at risk from unscrupulous vendors, and even an experienced gemstone buyer may make a mistake or be taken in by the atmosphere and what may seem like the "once in a lifetime" opportunity.

TRAVELING ABROAD

If you wish to see some of the best material that has been sourced locally or a range of specimens from around the world, including crystals, cut and polished gemstones that have not been set in jewelry, and jewelry (traditional and contemporary), you may wish to take an opportunity to visit local and national museums and galleries as well as jewelry centers and retail outlets. Be wary of street traders and market stalls. Short breaks to the capitals of Europe or longer trips further afield can reveal many hidden treasures.

This table shows just a few countries you might visit, the gem cutting and polishing areas and jewelry retail centers that you might like to head for, a few museums and galleries with royal regalia, collections of gemstones or jewelry, and some of the gemstones to look out for when visiting. The list is not exhaustive, and you may find your own favorite museums, visit workshops on your travels, see mineral and gem collections, and explore shops packed full of wonderful displays of minerals, gemstones, and jewelry that you would not find back home.

	GEM CUTTING AND POLISHING AREAS AND JEWELRY RETAIL CENTERS	MUSEUMS AND GALLERIES WITH COLLECTIONS OF GEMSTONES AND/OR JEWELRY	GEMS AND MINERALS, AND MINERAL LOCALITIES
Europe			
ENGLAND	The jewelry centers of London (Hatton Garden) and Birmingham (jewelry quarter).	The collection of royal regalia housed in the jewel house in the Tower of London. The Natural History Museum (London), Victoria and Albert Museum (London), Wallace Collection (London) have collections of minerals, gems, and jewelry.	Jet (Scarborough and Whitby), Blue John and other fluorspars, agates and jasper, etc. Peak District.
SCOTLAND		National Museum of Scotland (Edinburgh).	Smoky quartz and Cairngorm (Scotland).
WALES		Displays of Welsh gold at the National Museum of Wales (Cardiff).	Disused gold mines and slate mines (Wales).
BELGIUM	Antwerp is the jewelry center of Belgium. The diamond district houses diamond bourses and the Diamond Museum.	National Museum (Brussels). Royal Belgian Institute of Natural Sciences (Brussels). Diamond Museum (Antwerp).	
HOLLAND	Diamond-cutting factories (e.g. Stoeltie Diamonds, Gassen Diamond) offer guided tours (Amsterdam).	The Amsterdam Diamond Center.	
FINLAND		Ylarnaa Gem Museum.	Labradorite and iolite.
FRANCE	Gem shows in Paris in December.	The Louvre with crown jewels, etc (Paris). The Museum of Natural History and the Mineralogical Museum, housed in the University – École des Mines (Paris).	Tourist sites with lapidary museums for example in Avignon, Perigueux and Castlenay Castle.

	GEM CUTTING AND POLISHING AREAS AND JEWELRY RETAIL CENTERS	MUSEUMS AND GALLERIES WITH COLLECTIONS OF GEMSTONES AND/OR JEWELRY	GEMS AND MINERALS, AND MINERAL LOCALITIES
GERMANY	The center of the gem trade is in the joint towns of Idar-Oberstein.	The Edelstein Museum of Gemstones (Idar-Oberstein). Dresden's Green Vault houses the famous Dresden Green diamond. Museum Reich der Kristalle, Munich.	Area traditionally famous for agates, can still visit a traditional agate mine, see demonstrations of agate cutting (though most is imported) (Idar-Oberstein). Munich annual gem and mineral show.
CZECH REPUBLIC	Walk the lanes of Prague to see shops full of garnet jewelry.	Renowned for its Bohemian gems (garnets), visit the Natural History Museum (Prague)	Pyrope garnet.
ITALY	Fashion stores of Milan, Rome, Florence and Naples. Vicenza gem and jewelry fairs and trade shows.	Museums in Rome, Florence and Milan, see also the Castellani Collection of ancient Etruscan coins, gold and jewelry at the National Etruscan Museum in the Villa Giulia, Rome.	Coral workings in Naples and cameo workshops (coral, agate and shell)
SPAIN	If you are keen on making your own jewelry, there are a number of wholesale bead and fixtures traders who will sell to the individual (Barcelona and Madrid). Jewelry center (Cordoba).	Museo de Joyeria Regina jewelry museum (Cordoba). Mineralogical museum in Valverde de Camino and Seville. Museum of the School of Mines (Madrid).	Andalusite and chiastolite from Andalusia. Simulated pearls (Majorca). Peridot (Canary Islands).
SWITZERLAND	Center for precious stones (Geneva).	The Geological Museum of Lausanne.	
TURKEY		Topkapi Palace Museum (Istanbul).	Carved pipes made of meerschaum. Diaspore. Alexandrite, charoite, amber.
RUSSIA	Russia is a major supplier of synthetics.	The State Hermitage Museum and its amber room (St Petersburg). Fersman Mineralogical Museum, Moscow.	

THE MIDDLE EAST

SAUDI ARABIA, KUWAIT, BAHRAIN, DUBAI, QATAR, OMAN	Gold and silver rather than gemstones.	Pearl museum (Dubai) though pearl industry now mainly cultured.	Quartz pebbles marketed as "Saudi diamond."
ISRAEL	Major cutting center. Jewelry center in Ramat-Gan area of Tel-Aviv has Israel's Diamond Exchange and Colored Stone Exchange.	Geological Museum of the Hebrew University, Jerusalem.	Eilat stone cut *en cabochon*.

NORTH AMERICA

USA	Fashion stores of Fifth Avenue, New York. The Gem Museum/Store, (Honolulu, Hawaii) has more than 60,000 gems and minerals, including lapis from Afghanistan.	Smithsonian Institution (Washington, D.C.) houses the national collection of gemstones, including the Hope Diamond. American Museum of Natural History (New York). Harvard Mineral Museum, Boston.	Tuscon annual gem and mineral show, the largest in the world with more than 4,000 exhibitors. Turquoise used in traditional Native American jewelry.
CANADA		Diamond mining. Pacific Mineral Museum, Vancouver.	Ammolite and mammoth ivory (Rocky Mountains). Labradorite and sunstone. Nephrite jade.

	GEM CUTTING AND POLISHING AREAS AND JEWELRY RETAIL CENTERS	MUSEUMS AND GALLERIES WITH COLLECTIONS OF GEMSTONES AND/OR JEWELRY	GEMS AND MINERALS, AND MINERAL LOCALITIES
CENTRAL AND SOUTH AMERICA			
BRAZIL	Brazil supplies the world with a vast array of crystals and gems. Agates and geodes on sale in tourist shops (e.g. Rio Grande do Sul).	H. Stern Gem Museum, Rio de Janeiro, includes workshop tours and private collection of tourmalines, etc.	A must for the crystal and mineral collector.
MEXICO	Center of opal production and lapidary work is the state and city of Querataro, 134 miles (215 km) northwest of Mexico City.	Jade production and museums Antiqua Guatemala (Western Highlands) and San Cristóbal de las Cusas (Chiapas State).	Fire opal, water opal, jadeite and garnets.
DOMINICAN REPUBLIC	Amber sold in many shops.	Larimar Museum, Santo Domingo and amber museums.	Amber (Amber Valley). Blue pectolite, known as larimar.
AFRICA			
TUNISIA		The Bardo Museum (just outside Tunis) has Roman mosaics, artifacts and jewelry.	Compressed amber and amber simulants.
MADAGASCAR	Wide range of gemstones in dealers' shops and on market stalls.		Beryl.
ASIA			
CHINA	Wuzhou Gemstone City in Guangxi Province is China's largest costume jewelry center, covering a vast area of nearly 5.4 acres, with over half devoted to separate businesses and 500 shops.	Gem museums in university towns including Gem Museum in Dalian. Mineral museums including Mineral Museum, Shenzen (Guangdong Province) near the mouth of the Pearl River, Mineral Museum in Shekou, Nantau Peninsula	
HONG KONG	Large displays in commercial areas and large shopping malls.	Natural History Collection, Hong Kong Museum of History.	Natural, cultured and dyed pearls. Dyed jade.
INDIA			Moonstone.
INDONESIA	Market stalls may sell rough and faceted material including Java and Barjamasin.	Bandung Geological Museum, Bandung, West Java.	
SRI LANKA	Very wide range of gemstones, the "Land of gems." Ratnapura is the "City of gems."	Visit museums to see the range available, e.g. Gem Museum, Kandy.	Cat's-eyes of various gemstones. Moonstone.
THAILAND	Gemstone supermarkets (Bangkok) offering gemstones from around the world, particularly Thailand, Sri Lanka, India, Madagascar, and Burma (Myanmar). Major trading center for ruby and sapphire (Chantaburi), with weekend gem markets (for the professionals). Bangkok gem and jewelry show.	Museum of Mineral Resources, Bangkok.	
AUSTRALASIA			
AUSTRALIA			Opal, diamond and sapphire. Argyle Diamond Mine, famous for its pink diamonds.
NEW ZEALAND			Nephrite jade, paua shell and soapstone. Kauri gum may be mistaken for amber.

57

IMITATION AND SYNTHETIC GEMSTONES

▼ *Art Nouveau pendant set with an oval green paste (glass) gem and matching drop.*

Because of the value of gemstones and their rarity, there will always be a market for imitations or cheaper copies. Generally speaking, imitation gemstones are made to deceive. They are made from any material, natural or man-made, that looks similar enough to be taken for the real and more expensive gem.

Imitations have been made for at least 6,000 years, for example blue faience (glazed clay) was used by the Egyptians to imitate turquoise. Romans used colored glass to imitate gemstones including emeralds and rubies. The Victorians used a number of materials including glass (paste), plastics, and resins to imitate both the natural mineral gemstones and the organic gems such as amber and shell.

GLASS

Glass is a versatile material as it can be made almost any color and can be molded or faceted to give the required appearance. However, there are noticeable differences between glass and gems. Glass is generally not as hard as the gemstone that it imitates, and so will scratch more easily. Furthermore, its facet edges may appear scratched and worn. There may also be bubbles or distinctive swirls within the glass, which can be easily seen with the use of a hand lens (loupe) at a x10 magnification. A gemologist will also be able to distinguish glass by its single refraction and the value of its refractive index (1.5–1.7), as there are no singly refractive gems with the same refractive index.

DIAMOND SIMULANTS

One gemstone can be used to imitate another more expensive variety, for example citrine to imitate topaz, or colorless glass or quartz to imitate diamond. Colorless glass does not make a convincing diamond imitation (simulant) as it lacks the hardness, fire, and sparkle.

There have been other diamond simulants, including, for example, yttrium aluminum garnet (YAG) and strontium titanate, but all have either lacked fire (spinel, quartz, topaz, paste, plastics), or had too much (strontium titanate, rutile), been too soft or brittle. Diamond simulants can be distinguished from diamond by the fact that they conduct heat far less readily than diamond. Conventional diamond testers test for thermal conductivity and are useful in separating diamond from simulants such as cubic zirconia, corundum, glass, synthetic rutile, and zircon.

Other diamond imitations include CZ (cubic zirconia) and moissanite. Synthetic moissanite was produced in the 1970s, but early stones were yellowish. Moissonite is now produced as an almost colorless material which is nearly as hard as diamond, measuring more than nine on the Mohs scale and quite difficult to identify. The main difference is that diamond is singly refractive, while moissanite is doubly refractive. In larger moissonite gemstones, this can be seen as a doubling of the pavilion facets when viewed through the stone, but in small stones mounted in jewelry, moissanite can be difficult to identify. Thermal diamond testers cannot be used to discriminate diamond from moissanite, because both are thermally conductive. However, an electrical conductivity tester will indicate moissanite.

▲ *Cubic zirconia (CZ), a diamond simulant.*

COMPOSITES – garnet-topped doublets and soudé emeralds

Imitations may also be composites – made of more than one part. For example, a piece of green glass topped with a slice of red garnet can be used to imitate an emerald or green garnet. The garnet-topped doublet (GTD) is made of two parts and the junction between them can usually be recognized by the difference in luster. There may also be characteristic bubbles in the glass, which are not visible in the garnet.

When viewed through the table facet, the gemstone appears green, but when viewed from the side, or immersed in water, the red slice of garnet will be more obvious. GTD can be made to imitate gemstones of all colors by altering the color of the glass beneath the gemstone top. Another composite is the soudé emerald, made of two layers of colorless quartz, sandwiching a thin green layer of gelatine or glass.

▲ *Garnet-topped doublet (GTD), a composite imitation gem.*

COMPOSITES – opal doublets and opal triplets

Opal often occurs in thin seams with the precious opal forming just a thin sliver or slice. An opal doublet (made of two parts) is made by cementing a piece of precious opal (showing a play of color) to a "backing" layer of "potch" opal (non-precious opal). Opal doublets may also have a quartz, chalcedony, glass, or plastic base. Opal triplets have a protective covering of quartz above the opal in addition to a backing layer.

OPAL SIMULANTS

The play of color seen in precious opals is due to interference of light by its internal spherical structure. In 1974 the French scientist Pierre Gilson unveiled the first laboratory-produced opal. Gilson imitation opals can be distinguished from natural opals by their patchy appearance and the mosaic-like junctions between colored "grains." The US scientist John Slocum created a glass opal, known as "Slocum stone." The patches of color in Slocum stones have a slightly crumpled look when magnified.

SYNTHETICS

Gemstones can also be made in the laboratory. Where they have the same chemical composition as their natural counterpart, they are referred to as synthetic. Synthetic gemstones therefore have virtually the same physical and optical properties of the natural gemstone.

Verneuil flame-fusion method

In 1902, French scientist Auguste Verneuil (1856–1913) established the first commercially successful process for manufacturing synthetic rubies and sapphires. He used a flame-fusion process, melting the powdered ingredients (aluminum oxide and a coloring agent) and

▲ *Synthetic emerald,*
pendeloque cut.

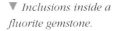

▼ *Inclusions inside a*
fluorite gemstone.

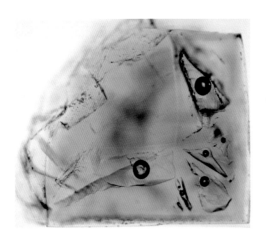

allowing these to drip on to a revolving stand. As the stand revolves, the drips cool and solidify, forming a candle-like shaped solid (boule) of the gem, which can then be faceted. Synthetic spinel can be made using the same process, and colored to imitate other natural gemstones, for example blue synthetic spinel to imitate aquamarine. Adding titanium oxide to the feed powder produces rutile inclusions, which form synthetic star stones. The rays of a synthetic star stone tend to look straighter and more regular than its natural counterpart.

Flux-melt and hydrothermal methods

In 1877 French chemist Edmond Fremy (1814–94) introduced the flux method of producing synthetic emeralds, which was improved in the 1930s and is still in use today. Ingredients are melted at high temperatures in a platinum crucible with a solvent (flux) before slowly cooling to produce the crystals. This flux-melt process is slower and more expensive than the Verneuil method, and is used for manufacturing synthetic emerald, ruby and alexandrite.

Inclusions in synthetic gemstones

A perfectly clear stone, without flaws or inclusions (internal features of gemstones, such as solids, liquids or gases, that have become enclosed within the gemstone during or after formation) is usually a warning sign to a gemologist that the stone may be synthetic. However, some synthetics may be identified by the type of inclusions, formed as a result of the method of manufacture. Flame-fusion gemstones may include specks of the powdered ingredients that have failed to melt sufficiently, and have curved growth-lines, rather than the straight lines that develop in natural specimens.

Synthetic-flux emeralds may have inclusions of the minerals phenakite or platinum (from the platinum crucible), twisted veils and feathers (characteristic patterns of inclusions), and two-phase inclusions (a liquid and a solid). Synthetic hydrothermal emeralds usually have only a few inclusions, possibly of phenakite and occasionally very fine two-phase inclusions.

Natural inclusions

Natural inclusions may be specific to one gem or even to a particular country or mine. Emeralds, for example, tend to be so heavily included that the view down a microscope (340 magnification) is sometimes referred to as *jardin* (French, "garden"). Natural emeralds may have inclusions of the minerals mica, tremolite, actinolite, pyrite, or calcite. Colombian emeralds may contain characteristic three-phase inclusions (solid, liquid, and gas), with a cubic salt crystal and a gas bubble within a fluid-filled cavity. Other gemstones with characteristic inclusions include peridot, which has inclusions that resemble "lily pads" (liquid droplets around a crystal of chromite) and the "treacly" appearance of hessonite garnet.

ENHANCEMENT

The color and appearance of natural and synthetic gemstones can be enhanced by methods such as oiling, dying, staining, foiling, heat-treating, irradiating, drilling, and filling. Heavily included gemstones, such as emerald, are oiled to fill cracks and give the stone a clearer appearance. Oiling has been carried out for more than 2,000 years, and is an acceptable part of the trade. Oiling, waxing, staining and filling cracks may be temporary, but more long-lasting resins have recently been introduced.

Oiling, staining, waxing and foiling

Oils and stains may leak and can be identified by wiping with a soft absorbent cloth. Turquoise, lapis lazuli, jade, and agates may be given a coating of wax. Agates are often stained, sometimes to bright colors that have no natural counterpart and are therefore easily recognized as fakes, though some find them attractive. Placing a piece of foil behind a mounted gemstone to increase its fire and sparkle was a method used by Victorians, particularly for costume jewelry (paste). The appearance or a closed mount may alert the gemologist to the possibility of foiling.

Heat-treatment

Probably the oldest method of enhancing the appearance of a gemstone is by heat-treating. In India, carnelian has been heat-treated for more than 4,000 years. Heating may be carried out in oxidizing conditions (with oxygen present) or in reducing conditions. For example, the pale creamy-gray sapphire from Sri Lanka, known as geuda, when heat-treated (2,730–3,450°F/1,500–1,900°C) in reducing conditions turns blue, and dark-colored rubies and sapphires may have their color lightened and improved by heat-treatment (1,470–2,550°F/800–1,400°C) in oxidizing conditions.

Heat-treatment and irradiation

Heat-treating and irradiating gemstones (with electrons or radioactive particles) can be used to change or enhance color. For example, colorless and pale brown topaz can be heat-treated or irradiated to give blue topaz. In addition, lasers can be used to drill even the hardest diamonds to remove inclusions and the drill hole may then be filled.

▼ *Heat-treated rubies (left) compared with natural rubies (right).*

High pressure, high temperature

High-pressure, high-temperature (HPHT) treatments have been used to improve the color of diamonds. The treatment has been used to produce red diamonds from natural diamonds of different colors. All red diamonds that have been produced by this treatment should have a mark on the girdle of the cut stone which has been inscribed by laser.

Laser drilling and filling

Diamonds that contain inclusions can be drilled to remove the inclusion and so increase the value of the gemstone. This leaves a permanent drill hole that may be seen as a flash of color when the diamond is moved in a light source. In order to make the drill hole less obvious it may be filled with a glass-like filling. This technique is not permanent and care should be taken not to dislodge the filling for example by using ultra-sound cleaning techniques.

Beryllium diffusion (bulk diffusion, surface diffusion)

The process was first used about 30 years ago. Following laboratory trials in 2002, gemologists studying pink-orange sapphires that were being sold in Thailand as padparadscha were able to report on the possible process that was being used. They were first alerted to the possible treatment because of the increase in padparadscha on the market without any notice of a new source.

They found that already faceted pink sapphires from Madagascar were being coated in borax and heated at high temperatures in the presence of oxygen (dissolving the outer layers and then reforming them in another color). An orange rim was formed that penetrated the surface or near-surface areas of the stone only. When removed from the furnace the orange coloration could be seen. The gemstones were then cleaned and repolished to reduce the obvious orange color, so that they look pink with only a subtle orange color (a padparadscha color).

The new technique is similar to other surface diffusion treatments that have been used for sapphire but penetrate further. The technique is now being used for most corundum varieties in order to enhance

▶ *Glass infill on a spinel is noticeable at this magnification.*

◀ *London blue topaz.*
"Sky blue," "Swiss blue" and
"London blue" are trade
names describing the color
of topaz. London blue topaz
is the darkest, usually due
to heat-treatment and/or
irradiation.

color including ruby and blue, green, yellow, orange and orange/red treated sapphires, as well as the original pink/orange treated stones. Green topaz, a color which does not occur naturally in topaz, is also becoming more common as a result of diffusion treatment.

DISCLOSURE

Most gemstones on the market have been enhanced or treated in some way. Some treatments such as oiling emeralds (to enhance clarity) or heat-treating sapphires (to enhance color) are so commonplace that they are seldom mentioned, others are easily identifiable. Many buyers are not aware that treatments and enhancements exist and some of the modern treatments are difficult to detect without gemological training and equipment. A jeweler should only refer to a gemstone as natural and untreated if it is just that. Disclosure of all treatments is encouraged, ideally without confusing or alarming the buyer, and all jewelers should be aware of what they are selling.

In the US, the obligation to fully disclose all treatments is covered by legislation. The US Federal Trade Commission (FTC) Guides for the gem and jewelry trade require disclosure of any treatments that substantially affect the value of the gems or jewelry. State deceptive practice regulations in the US require that vendors do not mislead customers about the treatment of the gems they sell. Non-disclosure can amount to fraud.

The World Jewelry Confederation (CIBJO) has introduced best-practice principles for its members to ensure the jewelry industry is run in an ethical and professional way and publishes rules, regulations and guidelines. Compliance includes full disclosure of all treatments to natural gemstones and of wholly or partly synthetic gemstones and of simulants at all levels of the jewelry distribution chain.

The history of gemstones is rich with stories of adventures, of legends, of tales of good fortune or curses resulting in financial ruin, ill health, or even death. Throughout antiquity, gemstones have been worn as a talisman, valued for their healing properties and endowed with spiritual values. The magical and mystical properties that are ascribed to gemstones have much to do with their rarity, beauty, feel, and color.

Gems have long been seen as exotic, rare and valuable. Many were carried long distances along perilous trade routes from distant and unknown lands, which added to their wonder and their value. As different gemstones have become available, fashions have changed and preferences have varied worldwide. Nowadays, a diamond is thought of as the ultimate gem, worn in engagement rings and given as a token of love, but this has not always been the case.

At various times in the past, TURQUOISE, AMETHYST, LAPIS LAZULI, JASPER, and CARNELIAN have all been regarded as the ultimate gem. Jade was a favorite in China and Mexico. The ancient Egyptians and the civilizations of Central and South America valued emeralds. Emeralds, sapphires, amethyst, jasper, and carnelian were the Romans' preferred choice, while diamonds were used to engrave cameos rather than be worn as jewelry.

Many of the stories associated with gemstones have been handed down by word of mouth. Others are gleaned from the diaries and letters of travelers or collectors, or as entries in the inventories of private collections, museums, or royalty. In the 13th century, Marco Polo (c. 1254–1324), a trader from Venice, traveled to Asia and wrote in his journal, *The Book of Marvels*, that he carried sapphires as calling cards when he visited the court of Kublai Khan, the Mongol Emperor. The sapphires were from southwest Sri Lanka (formerly Ceylon), from the area around Ratnapura ("City of Gems" in Sinhalese). In the 17th century, the French merchant Jean-Baptiste Tavernier (1605–89) made his fortune by trading in gemstones. Tavernier made six trips to India and Persia (now Iran) between 1631 and 1668, described many large diamonds and acquired a number of gems, some of which were sold to King Louis XIV of France.

Most of the famous, named gemstones are diamonds. Renaming and recutting as ownership changes may complicate their history, and the secrecy surrounding some gems and their whereabouts makes confirmation of size, shape, and weight difficult if not impossible. Museum specimens can be researched and some famous diamonds can be recognized from paintings or photographs, but those that are bought at auction

▲ *The Canning Jewel is named for the 2nd Viscount Canning, who bought it in c. 1860. Probably dating from the 16th century, the jewel is 4 inches (10 centimeters) long. The merman is made of gold and enamel, studded with rubies and diamonds, and his torso is formed by a large baroque pearl.*

▼ *This 14th-century manuscript illustration from the* Book of Marvels *shows Marco Polo being offered gems as he travels through Siam (present-day Thailand).*

by an "unknown private buyer," or those that are lost or are the victims of theft, simply "disappear," sometimes for many years. The azure "Nassak" (now 43 carats but originally 90 carats), also known as "The Eye of the Idol," was placed in the forehead of a statue of Shiva at a temple in Nassak, India, but disappeared when British troops looted the temple in 1818. In 1927 it resurfaced and was recut in New York.

The oldest diamonds with the longest histories largely originate from the alluvial deposits of the Golconda region of south-central India. They include the Koh-i-Noor, Orlov, Regent (Pitt), and Hope diamonds. Some of the largest and most famous diamonds are from the Premier Mine in South Africa, including the Cullinan and the Taylor-Burton (cut 69.42 carats). The largest diamonds in the world, the Golden Jubilee or Unnamed Brown (545.67 carats) and the De Beers Millennium Star (203 carats, which took 10 people two years to cut) are both African. In 1988 the Centenary diamond (599 carats uncut, 273.85 carats cut) was cut by the company De Beers to celebrate the 100th anniversary of its De Beers Consolidated mining operations.

Famous colored diamonds include the blue Hope diamond, the Dresden Green, and the golden-yellow Tiffany diamond (cut 128.54 carats). Other blue diamonds include the Townshend Blue (in the Victoria and Albert Museum, London), and the aforementioned pale blue Nassak. The Dresden Green is the world's largest pear-shaped green diamond (41 carats) and, apart from occasional loans, it has been kept in the vaults at Dresden Palace since its purchase by Frederick Augustus II of Saxony for US$150,000 at a Leipzig fair in 1743.

Today, pink diamonds from the Argyle mines in Western Australia are particularly prized. In 1986, a huge diamond referred to as the "Unnamed Brown" was used by De Beers to test their new laser cutting technology. The diamond weighed 755.50 carats when rough and 545.7 carats once cut. It was renamed the "Golden Jubilee" after its presentation to King Rama IX of Thailand in celebration of 50 years on the throne. Another brown diamond is the "Incomparable" (407 carats) found in the Democratic Republic of Congo (formerly Zaïre) in 1980. Black diamonds have been cut weighing more than 115 carats. The Black Orlov or "Eye of Brahma" cushion-cut black diamond (67.50 carats, not to be confused with the Orlov, see page 36) is said to have been stolen from a shrine in Pondicherry, southern India, and to have weighed 195 carats in the rough.

Other famous gemstones include the red spinels known as the Black Prince's ruby (in the British crown jewels) and the Kuwait ruby (formerly the Timur ruby), Saint Edward's sapphire and the Stuart sapphire (both in the British crown jewels), the Devonshire emerald, the Edwardes ruby, and the Rosser Reeves and Appalachian Star rubies.

▲ *The Centenary Diamond, weighing 273.85 carats, was cut to celebrate the 100th anniversary of the De Beers Consolidated mining operations. It is the largest modern-cut diamond in the world.*

▼ *The Koh-i-Noor diamond, 105.60 carats, at the center of the crown of Queen Elizabeth (the Queen Mother).*

HOPE DIAMOND

The blue Hope diamond, known largely for its history and legends of curses and bad luck, rather than its size (45.52 carats), is presently on display at the Smithsonian Institution in Washington, D.C. Named after its British buyer, Henry Philip Hope, there is strong evidence to suggest that the Hope diamond once formed part of the "Tavernier Blue." In 1642 Jean-Baptiste Tavernier bought a 116-carat diamond in India. In 1668 Tavernier sold the gem to King Louis XIV, who had it recut into a 67-carat stone and set in gold. In 1792, at the height of the French Revolution, the "French Blue" was stolen from the royal treasury.

KOH-I-NOOR DIAMOND

Koh-i-Noor means "Mountain of Light" in Persian. In legend, the Koh-i-Noor was worn by the god Krishna and the person who owns it is invincible. However, it is also said to bring misfortune to any male owner. The first record of the gem is in the memoirs of Babur, founder of the Mogul Empire in India. Babur recorded that the gem, weighing 739 carats, was among the treasures seized from the Raja of Malwa in 1304. The Mogul Emperors recut the stone, reducing it to 186 carats. It remained in Mogul hands until 1739, when the Persian ruler Nadir Shah conquered the Mogul Empire. In 1849 the British acquired the diamond, and the following year it was given to Queen Victoria. In 1852, the Amsterdam lapidary M. Coster recut the Koh-i-Noor to its present oval and weight of 105.60 carats. The recutting took nearly 450 hours of labor. In 1936 the Koh-i-Noor was mounted into the crown of Queen Elizabeth (the Queen Mother).

CULLINAN DIAMOND

The Cullinan is the largest gem diamond ever found. In 1905 Frederick Wells, a superintendant of the Premier Mine in South Africa, discovered the colossal gem embedded in a mine wall. According to reports, when Wells took it to be weighed, the inspector thought the stone was too large to be a diamond and threw it out of the window. Wells retrieved the gem, which was later authenticated at 3,106 carats – three times larger than any other known diamond. The diamond was named after the mine's owner, Colonel Thomas Cullinan.

The Transvaal government bought the stone and presented it to King Edward VII on his 66th birthday in 1907. In 1908 the Amsterdam lapidary Joseph Asscher was given the task of cleaving the diamond. Apparently, the knife broke on the first attempt but the diamond remained intact. A second cleavage yielded three main sections. Each day for eight months, diamond cutters worked on the stone, producing nine major gems (Cullinan I–Cullinan IX) and 96 smaller brilliants.

Cullinan I (also known as the "Star of Africa") was the largest diamond cut from the rough crystal. Cut as a pear-shape and weighing 530.20 carats, it is mounted in the Sovereign's Royal Scepter of the British crown jewels. Cullinan II is a 317.40-carat, cushion-cut stone mounted in the British imperial state crown.

Koh-i-Noor diamond,
105.60 carats

Cullinan I diamond,
530.20 carats

Cullinan II diamond,
317.40 carats

EUREKA DIAMOND AND THE STAR OF SOUTH AFRICA

Although not a particularly exceptional stone, the Eureka diamond (10.73 carats) has historic importance. The Eureka was cut from the first diamond to be discovered in South Africa. In 1866 Erasmus Jacobs, a young shepherd, picked up what he thought was a pebble on the banks of the Orange River, near Hopetown. The pebble was later discovered to be a 21.25-carat yellow diamond and prompted the diamond rush of 1867. The second diamond found in South Africa was the so-called "Star of South Africa." According to legend, the 82.5-carat rough stone was found in 1869 by a young shepherd, who exchanged it for 500 goats, 10 head of cattle, and a horse. It was cut into a 47.69-carat flawless pear-shaped diamond.

REGENT (PITT)

It is said that in 1701 a slave worker at an Indian mine discovered a 410-carat rough diamond. The slave escaped with the stone hidden inside a wound in his leg. In return for safe passage, the slave divulged his secret to an English sea captain, who murdered him. The captain then sold the diamond to an Indian merchant and soon after committed suicide. In 1702 the merchant sold the gem to Thomas Pitt, governor of Madras. Pitt had it cut into a 140.50-carat cushion-shaped brilliant cut. In 1717 it was sold to Philippe, Duc of Orleans and Regent of France, and henceforward was known as the "Regent diamond." In 1723 the diamond was part of the coronation crown of King Louis XV. In 1804, for his coronation as Emperor, Napoleon had it mounted in his sword hilt. In 1825 it was placed in the coronation crown of King Charles X. Emperor Napoleon III had it placed in a diadem for Empress Eugènie in 1889. It is now on display at the Louvre, Paris.

ORLOV

The size and shape of a pigeon's egg, the rose-cut Orlov is one of the world's biggest diamonds (189.62 carats) and is named after Count Orlov, who presented it to his lover Russian Tsarina Catherine the Great. Catherine mounted it in the Imperial Scepter, where it remains today on displayed in the Kremlin. In the 17th century, according to legend, the diamond was stolen from the eye of a statue of a Hindu god in a temple in Mysore, southern India. It was sold to an English sea captain for £2,000. It traveled via Persia to Amsterdam, where Orlov bought it in 1773.

SHAH

Unique in that it is engraved with the names of three shahs: "Bourhn-Nizam-Shah II, 1000" (Governor of India, 1591), "Son of Jehangir Shah Jehan Shah, 1051" (during the Government of Shah Jehan, 1641), and "Kadja Fath Ali Shah" (Shah of Persia). It was given to Tsar Nicholas I in 1829, and is displayed in the State Diamond Fund at the Kremlin.

▲ *Star of South Africa is a 47.69-carat flawless pear-shaped diamond. Its discovery prompted the Colonial Secretary to declare, "This diamond, gentlemen, is the rock upon which the future prosperity of South Africa will be built."*

▼ *Orlov diamond, 189.62 carats, has an upper surface marked by concentrated rows of triangular facets, with corresponding four-sided facets on the lower surface. In total, it has around 180 facets.*

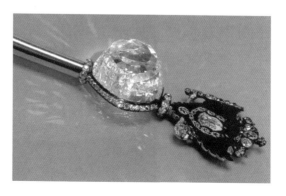

Gemstones and jewelry are bought to be worn and to be admired, so it is important to consider how best to look after your jewelry and gemstones. "Wear and tear" and dirt are the two main culprits; changes in heat and moisture can also affect gemstones.

WEAR AND TEAR

In order to reduce wear and tear to a minimum it is best to remove your jewelry when doing household work, washing the car, gardening, or taking part in physical activities or work. Harder gemstones such as diamond, ruby and sapphire are less likely to succumb to scratching, but may be cracked or broken if hit or knocked against a hard surface such as the edge of a table or sink. Avoid wearing several rings on the same or adjacent fingers where gemstones could scratch one another.

When removing jewelry each piece should be carefully stored. Ring trees and jewelry boxes help to keep pieces separate so that one gemstone does not scratch another.

DIRT AND GREASE

Dirt and grease on a gemstone can make an otherwise bright and sparkling piece of jewelry look dull. Household dust will settle on any gemstones that are not covered or protected. Try to avoid touching the gemstones. Dirt and grease can also collect at the back of a ring, caught in the setting or covering the back of the gemstone.

Pearls are sensitive to the chemicals in perfume and to the moisture on the skin. The acids wear away the outer layers of the pearl, removing the luster. Pearls are also sensitive to changes in heat, and as they are slightly porous, should be kept away from water as much as possible.

HEAT AND MOISTURE

The atmosphere or environment surrounding the gemstones may have an effect; for example, porous gemstones such as opal and turquoise may dry out in hot temperatures. Sudden temperature changes, such as sunbathing and then plunging into cold water or taking food out of a freezer, can cause a porous gemstone to crack. Porous gemstones should not be worn on aeroplane flights where the atmosphere is dry, or moved from dry to moist areas rapidly.

CLEANING GEMSTONES AND JEWELRY

A soft brush such as an old toothbrush can be used to clean between the prongs of the setting and also to clean the back of the stone, unless in a closed setting where the back is not accessible. Be careful not to damage settings as this may cause the gemstones to become loose and possibly dislodged. Avoid using cloths or towels (except possibly soft jewelry cloths) to remove dust or dry jewelry, as the prongs of the settings could become entangled in the threads of the material and damage the settings.

Diamonds may be cleaned using a small amount of warm soapy water and a soft brush. Do not be tempted to scratch dirt off: just leave it a little longer and try again with the soft brush or jewelry

▼ Clean jewelry carefully using a soft brush, such as an old toothbrush.

cloth. Rinse in warm water and leave to dry. The same method can be used with most colored gemstones.

Emeralds are usually oiled and should not be immersed in hot water as this could cause fillings to leak and the cracks to become more obvious. Do not soak in cleaning solutions as this could also dissolve fillings. Ultra-sound cleaners should not be used with emerald jewelry as cracks may be made worse and the gemstone might break. Diamonds, rubies and sapphires are generally fine in ultra-sound cleaners, but check that rubies are not heavily included.

Opals and turquoise should not be cleaned with hot or cold water or with washing-up liquid, but can be wiped with a gentle cloth. Ultra-sound cleaners should be avoided as opals and turquoise are porous and the ultra-sound bath could damage the gemstones. Pearls may be wiped with warm soapy water but should not be left to soak. Jewelry is particularly susceptible to damage from the effects of chemicals present in perfume, hairspray, nail polish remover and deodorant.

▼ *A cleaning guide table to the more commonly found gemstones. If you have any doubt how best to care for your particular piece of jewelry visit your local jeweler, who should be able to advise you on the best course of action.*

CLEANING GUIDE		
	Soak in warm soapy water	Use ultra-sound cleaner	Notes
AMBER	X	X	Avoid strong heat, strong sunlight, pressure, acids and chemicals (solvents, etc.).
AMETHYST	✓	✓	Avoid strong heat, abrupt temperature changes.
AQUAMARINE	✓	✓ with caution	Avoid strong sunlight.
CHRYSOBERYL	✓	✓ with caution	
CITRINE	✓	✓ with caution	Avoid strong heat, abrupt temperature changes, strong sunlight.
DIAMOND	✓	✓	
EMERALD	X	X	Avoid strong sunlight.
GARNET	✓	✓ with caution	Avoid abrupt temperature changes.
JADEITE AND NEPHRITE JADE	✓	✓ with caution	Avoid strong heat, strong sunlight, acids and chemicals.
OPAL	X	X	Avoid acids and chemicals, strong heat, strong sunlight.
PEARL	X	X	Avoid acids and chemicals, strong sunlight.
PERIDOT	✓	X	Avoid acids and chemicals, strong sunlight.
RUBY	✓	✓ with caution	
SAPPHIRE	✓	✓ with caution	
SPINEL	✓	✓ with caution	Avoid strong heat.
TANZANITE	✓ with caution	X	Avoid abrupt temperature change, acids and chemicals, strong sunlight.
TOPAZ	✓ with caution	✓ with caution	Avoid strong sunlight.
TOURMALINE	✓	✓ with caution	Avoid abrupt temperature change.
ZIRCON	✓ with caution	✓	Avoid strong sunlight.

COLOR KEY

Color can be a useful clue to aid in identification of a gem, but it can also mislead, so beware! Gemstones may look different under different light sources; for example, chrysoberyl with a color change may show flashes of green in daylight, but has flashes of red indoors under artificial light. The color of a gemstone may be enhanced or altered (see pages 61–3), and imitations (see pages 58–60) may be fashioned to look exactly the same color as the natural gem they imitate.

A gemologist will take the color of the gemstone into account, but will also look for other clues; for example, wear and tear of the facet edges, scratches and pit marks give an idea of hardness, while fire and sparkle suggest possible values for dispersion and refractive index. Appropriate tests may then need to be carried out to establish the identification of the gemstone.

YELLOW TO BROWN GEMS

AGATE
see pages 202-3

AMBER
see page 233

AMBLYGONITE
see page 120

ANGLESITE
see page 116

ARAGONITE
see page 108

AXINITE
see page 152

BARYTE
see page 115

BRAZILIANITE
see page 119

CALCITE
see page 104

CARNELIAN
see page 196

CASSITERITE
see page 100

CHATOYANT QUARTZ
see page 192

CHRYSOBERYL
see page 90

CITRINE
see page 188

DIAMOND
see pages 78-81

DRAVITE
see page 167

YELLOW TO BROWN GEMS (CONTINUED)

ENSTATITE
see page 170

EPIDOTE
see page 150

FIRE OPAL
see pages 205-6

HELIODOR
see page 158

HESSONITE
see page 137

HYPERSTHENE
see page 171

ORTHOCLASE
see page 208

PADPARADSCHA
see page 96

PREHNITE
see page 183

PYRITE
see page 83

RUTILE
see page 101

SAPPHIRE
see pages 93-4

SARD
see page 198

SARDONYX
see page 200

SINHALITE
see page 113

SMOKY QUARTZ
see page 190

SPHALERITE
see page 82

SPHENE
see page 129

SUNSTONE
see page 213

TOPAZ
see pages 146-8

TORTOISESHELL
see page 229

ZIRCON
see pages 130-1

OTHER YELLOW TO BROWN GEMS INCLUDE:

Apatite	Spessartine
Fluorite	garnet
Scheelite	Tourmaline
Staurolite	Vesuvianite

RED TO PINK GEMS

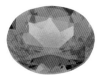

ALMANDINE
see page 135

CORAL
see pages 230-1

CUPRITE
see page 85

GROSSULAR GARNET
see page 138

JASPER
see page 195

LEPIDOLITE
see page 181

MORGANITE
see page 160

PYROPE
see page 134

RED BERYL
see page 161

RHODOCHROSITE
see page 105

RHODONITE
see page 178

ROSE QUARTZ
see page 189

RUBELLITE
see page 164

RUBY
see pages 97-9

SAPPHIRE
see page 93-5

SCAPOLITE
see page 218

SPESSARTINE
see page 136

SPINEL
see pages 86-7

SPODUMENE (KUNZITE)
see page 177

WATERMELON
TOURMALINE
see page 163

ZIRCON
see pages 130-1

ZOISITE (THULITE)
see page 151

OTHER RED TO PINK GEMS INCLUDE:

Diamond (rarely)	Rutile
Garnet-topped doublet	Smithsonite (pink)
	Topaz
Violet jadeite	Tourmaline

BLUE TO VIOLET GEMS

AMETHYST
see pages 186-7

AQUAMARINE
see pages 156-7

AZURITE
see page 112

DIAMOND
see pages 78-81

DUMOURTIERITE
see page 132

FLUORITE
see pages 102-3

HAÜYNE
see page 215

INDICOLITE
see page 165

IOLITE
see page 162

KYANITE
see page 144

LAPIS LAZULI
see pages 216-7

LAZULITE
see page 124

PECTOLITE
see page 178

SAPPHIRE
see pages 93-5

SILLIMANITE
see page 143

SMITHSONITE
see page 106

SODALITE
see page 214

SPINEL
see pages 86-7

TOPAZ
see pages 146-8

TURQUOISE
see pages 122-3

ZIRCON
see pages 130-1

ZOISITE (TANZANITE)
see page 151

OTHER BLUE TO VIOLET GEMS
INCLUDE:

Agate	Axinite
(stained)	Euclase
Apatite	Scapolite

GREEN GEMS

ALEXANDRITE
see page 91

ANDASLUSITE
see page 142

ANDRADITE
see page 139

APATITE
see page 121

BLOODSTONE
see page 197

CHRYSOCHOLLA
see page 111

CHRYSOPRASE
see page 201

DEMANTOID GARNET
see page 140

DIOPSIDE
see page 172

DIOPTASE
see page 128

EMERALD
see pages 154-5

HIDDENITE
see page 177

JADEITE
see pages 174-5

KORNERUPINE
see page 127

MALACHITE
see page 110

MICROCLINE
see page 210

NEPHRITE
see page 176

PERIDOT
see page 125

SAPPHIRE
see pages 93-5

SERPENTINE
see page 182

TEKTITE (MOLDAVITE)
see pages 222-3

TOPAZ
see pages 146-8

TSAVORITE GARNET
see page 138

UVAROVITE GARNET
see page 141

GREEN GEMS (CONTINUED)

VESUVIANITE
see page 169

ZIRCON
see pages 130-1

OTHER GREEN GEMS INCLUDE:

Agate	Sapphire
Diamond (rarely)	Tourmaline
Enstatite	Watermelon
Euclase	tourmaline
Fluorite	(pink and
Prehnite	green)

COLORLESS GEMS

ACHROITE
see page 166

CELESTINE
see page 114

CERUSSITE
see page 109

DANBURITE
see page 173

DIAMOND
see pages 78-81

DOLOMITE
see page 107

EUCLASE
see page 149

GOSHENITE
see page 159

NATROLITE
see page 219

PHENAKITE
see page 126

ROCK CRYSTAL
see page 185

SCHEELITE
see page 118

TOPAZ
see pages 146-8

ZIRCON
see pages 130-1

OTHER COLORLESS GEMS INCLUDE:

Albite	Grossular garnet
Anglesite	Moonstone
Apatite	Orthoclase
Calcite	Petalite
Enstatite	Sapphire
Fluorite	Scapolite

WHITE GEMS

GYPSUM
see page 117

MILKY QUARTZ
see page 191

MOONSTONE
see page 209

OPAL
see pages 205-6

PEARL
see page 227

SHELL
see page 228

OTHER WHITE GEMS INCLUDE:

Agate	Jadeite
Baryte	Nephrite
Calcite	Serpentine
Coral	Thomsonite
Ivory	

BLACK GEMS

DIAMOND
see pages 78-81

HEMATITE
see page 88

JET
see page 236

OBSIDIAN
see page 221

ONYX
see page 199

SCHORL
see page 168

TEKTITE
see pages 222-3

OTHER BLACK GEMS INCLUDE:

Andradite
(melanite)
Coral
Pearl

IRIDESCENT GEMS

FIRE AGATE
see page 204

LABRADORITE
see page 212

MOTHER-OF-PEARL
see page 228

OPAL
see pages 205-6

In the gem-description section that follows, we have organized the gems by their chemical composition. The gems are grouped in the following classification scheme:

CARBON (diamond, pages 78–81)

SULFIDES (pages 82–4)

OXIDES AND HYDROXIDES (pages 85–101)

HALIDES (pages 102–3)

CARBONATES (pages 104–112)

BORATE (sinhalite, page 113)

SULFATES AND CHROMATES (pages 114–17)

TUNGSTATE (scheelite, page 118)

PHOSPHATES (pages 119–124)

SILICATES (pages 125–220)

IGNEOUS ROCK (obsidian, page 221)

TEKTITES (pages 222–3)

SYNTHETICS AND IMITATIONS (pages 224–5)

ORGANICS (pages 226–36)

At the end of the main section, we describe the precious metals (gold, silver, platinum, pages 237–43).

This scheme has the advantage of placing each gem species within its mineralogical class, enabling the reader to see an individual gem in relation to its nearest chemical associates.

When appropriate, we have placed gem species in mineral groups within each mineralogical class. For instance, sapphire is a gem species of the corundum group within the oxide and hydroxide class of minerals.

The silicates are such a large class that, while the primary classification is based on a chemical criterion, they are subdivided on a structural basis.

The tables on this page list the chemical symbols of the elements referred to in the book.

CHEMICAL SYMBOLS	
AG	SILVER
AL	ALUMINUM
AU	GOLD
B	BORON
BA	BARIUM
BE	BERYLLIUM
C	CARBON
CA	CALCIUM
CL	CHLORINE
CR	CHROMIUM
CU	COPPER
F	FLUORINE
FE	IRON
H	HYDROGEN
K	POTASSIUM
LI	LITHIUM
MG	MAGNESIUM
MN	MANGANESE
NA	SODIUM
O	OXYGEN
P	PHOSPHORUS
PB	LEAD
S	SULFUR
SI	SILICON
SN	TIN
SR	STRONTIUM
TI	TITANIUM
W	TUNGSTEN
ZN	ZINC
ZR	ZIRCONIUM

ANIONIC GROUPS	
Al_2O_4 etc.	ALUMINATE
BO_3, B_3O_4 etc.	BORATE
Cl, Cl_2 etc.	CHLORIDE
CO_3	CARBONATE
CrO_4 etc.	CHROMATE
F, F_2 etc.	FLUORIDE
O, O_2 etc.	OXIDE
$OH, (OH)_2$ etc.	HYDROXIDE
PO_4 etc.	PHOSPHATE
S, S_2 etc.	SULFIDE
SiO_4, Si_2O_7 etc.	SILICATE
SO_4	SULFATE
WO_4 etc.	TUNGSTATE

iamond is the hardest natural substance and the most valued gemstone. The name comes from the Greek word *adamas* meaning "invincible," and this alludes to the hardness and durability of diamond. The stones are more intensively mined and strictly graded than any other gemstone. Diamond is composed of pure carbon, like graphite, but its extreme hardness is a result of its atoms being compacted and bonded by high pressures and high temperatures in the Earth's upper mantle. Most diamonds are formed at depths of between 50 and 90 miles (80 and 150 kilometers) in the Earth's crust. Because diamond is made up of carbon, it burns in oxygen or in air heated to a very high temperature. The idea that diamond is combustible was first proposed in 1675 by Sir Isaac Newton (1642–1727), the English scientist and mathematician, and the first incineration of diamond was completed 19 years later by two Italians, Averani and Targioni.

Diamond commonly occurs as octahedral crystals frequently of flattened habit, and more rarely as cubes, often with curved faces. Most diamonds came from alluvial deposits (mainly river and beach gravels) until the discovery in the mid-19th century in South Africa of pipe-like intrusions that have risen from great depths and are filled with a variety of peridotite that was named **kimberlite**, being first identified in Kimberley. This igneous rock is rich in magnesium, iron, and calcium. Such intrusions are also found in lamproitic rocks, which are the source of diamonds in Western Australia. Intrusions can remain deep in the Earth until volcanic eruptions force them to the surface within hours. Some diamonds have also been found in iron meteorites.

▲ *Round brilliant-cut diamond, 17.16 carats (top); brilliant-cut diamond, 5.25 carats (middle); cushion-cut yellow diamond (bottom).*

Diamonds were known in India some 2,300 years ago but were not deliberately broken or cut, since it was thought this would destroy magical properties of the stone. Roman scholar and naturalist Pliny the Elder (AD 23–79) recorded the existence of diamonds in the 1st century, but they had little value because

DIAMOND	
CHEMICAL COMPOSITION	C
COLOR	COLORLESS, YELLOWISH, BROWN, RED, BLACK
REFRACTIVE INDEX	2.41
RELATIVE DENSITY	3.5
HARDNESS	10
CRYSTAL GROUP	CUBIC
CLEAVAGE	PERFECT
FRACTURE	CONCHOIDAL
TENACITY	NONE
LUSTER	ADAMANTINE
TRANSPARENCY	TRANSPARENT TO TRANSLUCENT
DISPERSION	STRONG, .044
BIREFRINGENCE	NONE
PLEOCHROISM	NONE

LOCATION Angola, Australia, Borneo, Botswana, Brazil, Canada, China, Democratic Republic of Congo, Ghana, Guinea, Guyana, India, Namibia, Russia, Sierra Leone, South Africa, Tanzania, United States, Venezuela, Zimbabwe.

▶ *Rough diamonds such as these are found in pipe-like intrusions in kimberlite and lamproitic rocks and in alluvial deposits.*

the Romans could not solve how to cut such a hard stone. In the 14th century, Europeans began to produce table cuts from octahedral crystals and rose cuts from cleavage fragments. The brilliant cut first appeared in the 17th century and was greatly improved in 1919. Lasers are now used to cut diamonds.

Diamonds are the primary gem measured in carats. The value of a carat has been different in various countries through the years, but has been standardized as 0.2 grams or 200 milligrams since 1907. The term "carat" supposedly derives from the weight of Mediterranean carob seed pods, which have long been used to weigh precious stones. The name is derived from the Greek term *keration*, meaning "fruit of the carob."

Diamonds have a perfect crystal form and high symmetry. Uncut crystals may look greasy and somewhat rounded, but when broken or cut, the faces show brilliant adamantine (diamond-like) luster and dispersion. This gives the stone a fiery brilliance that is probably best displayed by brilliant cuts, which are the most popular today. This cut reflects the most light possible through the front of the gem. Other common cuts are the cushion, pendeloque, marquise, fancy, round, and square cuts. Diamonds are popular for all types of jewelry, such as earrings, brooches, rings, and bracelets. The stones are traditionally set in gold and platinum. "Meleé" is the term used to describe a group of small diamonds, each weighing less than 0.25 carats, that are used to embellish the mountings for larger gems.

DIAMOND

Major diamond fields are located in Kimberley, Transvaal, South Africa, with the first diamond being reported in 1866 in gravels near the River Orange. Famous diamonds, such as the Koh-i-Noor, have come from mines in the Golconda area of southern India. India was the major source of diamonds until about 1725, when diamonds were discovered in Brazil. Small stones of good quality are still found in Brazil. Since the late 19th century, South Africa has been the major source of gem-quality diamonds. Australia is a major producer of fancy colors. The Democratic Republic of Congo (formerly Zaïre) once produced the largest quantity of diamonds, almost all being of industrial grade. Diamonds also exist in the United States in Pike County, Arkansas, and the Kelsey Lake Mine, Colorado. Alluvial diamonds have been discovered in nearly every US state. Canadian stones are found on St Helen's Island in Montréal, Québec, and in the Northwest Territories. Other major fields exist in Serbia, Russia, Venezuela, and from the seabed off Namibia.

Mouawad Splendor,
101.84 carats

Mouawad Magic,
108.81 carats

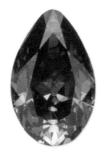

Mouawad Blue,
42.92 carats

The Ahmedabad,
78.86 carats

Brilliant

Cushion

Pendeloque

Fancy

Marquise

The gem-quality specimens, which are sufficiently transparent, colorless and unflawed to be used in jewelry, for example, account for less than one-quarter of mined diamonds. More than half of their weight is lost during cutting. The value of diamonds varies according to their color and transparency. Perfectly colorless and transparent gems are the most expensive and rare, since most diamonds have impurities, usually nitrogen that causes yellowish, brown, green, and black colors. The rarest colors are red, blue, and green. Deep shades are called "fancy colors" and are greatly valued; the most sought after is blood red. Diamonds are also colored artificially using irradiation.

Because of its hardness, the only stone that can cut and polish a diamond is another diamond. **Imitation** diamonds have been made using SPINEL, SAPPHIRE, ZIRCON, strontium titanate, rock crystal, glass, paste, the man-made YAG (yttrium aluminum garnet), and other stones and materials. Synthetic diamonds were first produced in 1955, and are widely used for industrial purposes.

▲ *Round brilliant-cut black diamond ring, 16.56 carats, flanked by two pear-shaped diamonds.*

▶ *This Belle Epoque diamond necklace (c. 1905) has a beautiful independent old European-cut yellow diamond at its center.*

More than three-quarters of all mined diamonds are of industrial quality, although this percentage varies from mine to mine. Usually colored gray to brown, these finely granular diamonds are called bort, ballas, or carbonado. They are low-quality stones used in drill bits, glass cutters, styli for record players, saws for cutting building stones, and dies that draw fine wire. Other uses include fine-grinding and polishing, as well as for cutting other diamonds. One diamond can be used to cut another because a crystal will be more brittle in certain directions than in others. The ease of cleavage can also be used to split crystals because diamonds have perfect cleavage in four different directions that are parallel to the octahedral crystal faces.

Diamonds are classified as Type I or Type II according to their physical properties. Type I stones contain nitrogen, and these are further divided into Type Ia, which have the nitrogen in layers, and Type Ib, which have the nitrogen dispersed. About 999 out of every 1,000 diamonds are Type I crystals. Type II diamonds, such as naturally blue ones, have no nitrogen and are laminated. Type IIa stones do not conduct electricity, but Type IIb will. Fine specimens are found in both Type I and Type II, and some diamonds are a mixture of the types.

The largest gem diamond so far discovered is the Cullinan (see page 35), which weighed 3,106 carats. It was found in 1905 at the Premier Diamond Mine near Pretoria in the Transvaal, South Africa, and later cut into brilliants as nine large stones and 96 others. The largest of the cut stones is "Cullinan I," also called the "Star of Africa" (530.20 carats), is mounted in the Sovereign's Royal Scepter of the British crown jewels. Another diamond, the blue Hope (45.52 carats) from India, is infamous for its reputation of bringing bad luck to anyone who owns it. The Hope (see page 34) is now on display in the Smithsonian Institution, Washington, D.C. Also on display in the Smithsonian is the largest known crystal that is perfectly developed, the yellow "Oppenheimer" (253.70 carats) from South Africa. The largest industrial diamond, a mass of carbonado (3,167 carats), came from the Brazilian state of Bahia.

◄ *Brooch with circular- and baguette-cut diamonds.*

◄ *Round brilliant-cut treated yellow diamond, 10.12 carats, surrounded by a ring of marquise-cut diamonds.*

▼ *The Cullinan I or "Star of Africa," seen here in the Sovereign's Royal Scepter of the British crown jewels, is a pear-cut diamond weighing 530.20 carats.*

SPHALERITE

Brilliant

Mixed

Sphalerite's name gives a clue to its nature – it derives from the Greek word *sphaleros* meaning "misleading.". Since it is variable in color and appearance, it can be hard to identify and is often confused with other minerals. Over the years, it has attracted many other names such as: "blende" from the German word for "blind or deceiving"; black jack; and "mock lead-ore."

Although it is too fragile for jewelry, sphalerite is a magnificent stone for collectors. Its high refraction and high dispersion give it a fire higher than a diamond. It also looks wonderful in a display with other gemstones.

SPHALERITE

Sphalerite is an important zinc ore and is exploited industrially. It is frequently associated with galena in hydrothermal veins and also occurs in limestones where ore bodies have been formed by replacement.

Spalerite is one of the few minerals that has six directions of cleavage. This makes cutting very difficult. Transparent stones suitable for cutting are found at Santander, northern Spain, and Naica, Chihuahua, northern Mexico.

▲ *Brilliant-cut sphalerite gem.*

◄ *Oval brilliant-cut sphalerite gem.*

SPHALERITE	
CHEMICAL COMPOSITION	ZnS
COLOR	DARK BROWN TO BLACK, SOMETIMES GREEN, ORANGE-RED OR YELLOWISH-BROWN
REFRACTIVE INDEX	2.37–2.42
RELATIVE DENSITY	3.9–4.2
HARDNESS	3.5–4.0
CRYSTAL GROUP	CUBIC
CLEAVAGE	PERFECT
FRACTURE	CONCHOIDAL
TENACITY	BRITTLE
LUSTER	METALLIC TO VITREOUS
TRANSPARENCY	TRANSPARENT TO TRANSLUCENT
DISPERSION	VERY STRONG, .156
BIREFRINGENCE	NONE
PLEOCHROISM	NONE
LOCATION Australia, Burma (Myanmar), Germany, Mexico, Morocco, Peru, Spain, United Kingdom, United States.	

▲ *Sphalerite crystals on matrix.*

Pyrite is one of the most common and widely distributed sulfide minerals. Although lighter than gold, prospectors used to mistake pyrite for the precious metal, earning it the synonym "fool's gold." It also resembles MARCASITE. It can be distingushed from gold and marcasite by its crystal form. Its name comes from the Greek word *pyr* meaning "fire" because pyrite emits sparks when struck. It has been used to start fires since prehistoric times, and is now sometimes used to produce sulfur, sulfuric acid, and iron sulfate.

When found in its raw state, pyrite crystals can be shaped as cubes, octahedrons, and pyritohedrons (12 faces). Twinning causes "iron crosses" that look like interpenetrating cubes. Pyrite is highly reflective and the Incas used pyrite tablets as mirrors. The ancient Greeks polished pyrite to make pins, earrings, and amulets. Pyrite gemstones were popular in Victorian Britain, but today are mostly used for costume jewelry, and are often fashioned into cabochons. Careful cutting is required because of its brittleness. Collectors like a flattened nodular variety called "pyrite suns" or "pyrite dollars."

Cabochon

Polished

PYRITE

Pyrite is present in igneous rocks as an accessory mineral, in sedimentary rocks, especially black shales, and in metamorphic rocks, notably in slates when it frequently forms well-shaped cubic crystals. Pyrite is a common replacement mineral in fossils, where it will take the place of weaker minerals present in certain bone and shell materials.

Among the richest deposits are those at Rio Tinto, southwest Spain. Large crystals exist on the island of Elba, west Italy, and high-quality specimens are found near Freiberg, southwest Germany. Pyritized fossils have been found worldwide. Pyrite is mined in many US states, including Vermont, New York, and Pennsylvania. It is also found in Australia, South Africa, Switzerland, Italy, China, Czech Republic, Sweden, Greece, Norway, Russia, Japan, and Peru.

PYRITE	
CHEMICAL COMPOSITION	FeS_2
COLOR	PALE BRASS-YELLOW
REFRACTIVE INDEX	NONE
RELATIVE DENSITY	4.9–5.2
HARDNESS	6–6.5
CRYSTAL GROUP	CUBIC
CLEAVAGE	INDISTINCT
FRACTURE	CONCHOIDAL TO UNEVEN
TENACITY	BRITTLE
LUSTER	METALLIC
TRANSPARENCY	OPAQUE
DISPERSION	—
BIREFRINGENCE	NONE
PLEOCHROISM	NONE

LOCATION Australia, China, Czech Rep., France, Germany, Greece, Italy, Japan, Mexico, Norway, Peru, Russia, South Africa, Spain, Sweden, Switzerland, United Kingdom, United States.

▲ *Mass of inter-penetrating cubic pyrite crystals.*

◄ *Interpenetrating cubes of pyrite.*

MARCASITE

Rose cut

An iron sulfide that is common worldwide. Marcasite is a polymorph of pyrite, having the same chemistry but different structures and crystal shapes. It was confused with pyrite until a French mineralogist, R. J. Haüy (1743–1822), distinguished it in 1814. Marcasite and pyrite are very similar in appearance, but marcasite has a paler color (causing it to be called "white iron pyrite"). Both were used as medicines in the Middle Ages. Its name derives from an Arabic or Moorish word used for pyrite and other minerals like bismuth and antimony.

MARCASITE

Marcasite is deposited at low temperatures in hydrothermal veins containing zinc and lead ores. It is found in near-surface deposits, most commonly in sedimentary rocks such as limestone, especially chalk, and clay, as single crystals, concretions, or as a replacement mineral in fossils. Exposed to air, it oxidizes easily and finally crumbles to dust. Over decades, marcasite in collections will disintegrate, freeing sulfur that forms sulfuric acid, which eats into the specimens' cardboard boxes.

Deposits have been discovered at Dover and in Derbyshire (spearhead-shaped form), England. In the United States, marcasite is mined in Hardin County, Illinois; Grant County, Wisconsin; and at Joplin, Missouri. Other significant sources include Carlsbad and Rammelsberg, Germany, as well as at sites in the Czech Republic, Russia, China, France, Romania, Peru, and Mexico.

The crystals are commonly tabular, often with curved faces, but also massive stalactitic or as radiating fibers. Twinning is common and can produce a spearhead-shaped form (sometimes mistaken for Roman weapons) or "cockscomb aggregates." In Victorian times, marcasite was popular for jewelry, and today it is often fashioned into reasonably priced rings, earrings, pendants, and bracelets. Small rose cuts are common. Confusingly, jewelers still use the name marcasite in reference to small polished and faceted pyrite stones inlayed in sterling SILVER. Polished marcasite has also been used to produce ornaments.

▲ *Twinned marcasite crystals forming "cockscombs."*

▶ *Interpenetrating cubes of marcasite, 4 carats.*

MARCASITE	
CHEMICAL COMPOSITION	FeS$_2$
COLOR	PALE BRONZE-YELLOW
REFRACTIVE INDEX	1.8
RELATIVE DENSITY	4.8–4.9
HARDNESS	6–6.5
CRYSTAL GROUP	ORTHORHOMBIC
CLEAVAGE	POOR
FRACTURE	UNEVEN
TENACITY	BRITTLE
LUSTER	METALLIC
TRANSPARENCY	OPAQUE
DISPERSION	——
BIREFRINGENCE	——
PLEOCHROISM	NONE
LOCATION China, Czech Republic, England, France, Germany, Mexico, Peru, Romania, Russia, United States.	

▲ *Cushion-cut cuprite gem.*

Sometimes known as "ruby copper" because of its deep red color, cuprite is a major ore of copper. Although it is mined throughout the world and is commonly used industrially, cuprite is comparatively rare as a gemstone. The name derives from *cuprum*, the Latin word for "copper."

For the gem collector, the coloration of cuprite is appealing. In the best samples, true deep reds are visible in reflection inside the almost black crystal. Its color is very similar to PYROPE garnet. Cuprite has a high refractive index, greater even than diamond. It also has a semimetallic luster and can easily be mistaken for HEMATITE.

Well-developed cubic crystal forms are the most sought after, and good crystals are rare. Although it is very attractive, cuprite is too soft to be used as jewelry, and it is extremely difficult to cut. Only some crystals found in Santa Rita, New Mexico, United States, and Onganja, Seeis, central Namibia, are suitable for cutting into brilliants for display.

Chalcotrichite, a variety of cuprite, is also attractive to collectors. Consisting of fine, red, hair-like crystals of cuprite, chalcotrichite samples have a very high sparkle. Its name literally means "hairy copper."

Brilliant

Cushion

CUPRITE	
CHEMICAL COMPOSITION	Cu_2O
COLOR	CRIMSON-RED
REFRACTIVE INDEX	2.85
RELATIVE DENSITY	6.1
HARDNESS	3.5–4
CRYSTAL GROUP	ISOMETRIC-HEXOCTAHEDRAL
CLEAVAGE	NONE
FRACTURE	CONCHOIDAL
TENACITY	BRITTLE
LUSTER	ADAMANTINE OR SUBMETALLIC
TRANSPARENCY	TRANSPARENT TO TRANSLUCENT
DISPERSION	————
BIREFRINGENCE	NONE (ISOTROPIC)
PLEOCHROISM	————

LOCATION Worldwide, especially Chile, Namibia, United States.

CUPRITE

Cuprite is usually formed as a secondary mineral in the oxidized zone of copper deposits, and it is commonly accompanied by MALACHITE, AZURITE and chalcocite. Small deposits of cuprite (copper oxide) are sometimes found in association with malachite or azurite on ancient coins that have been buried for thousands of years.

Major sources of cuprite include the Flinders Range, South Australia; Atacama Desert in Chile; Haut-Rhin, Alsace, northeast France; the Black Forest, Baden Württemberg, southwest Germany; Lavrion, Attica, east-central Greece; Zacatecas, north-central Mexico; and Cornwall, southwest England.

▲ *Cuprite crystals on matrix.*

For centuries, spinels have been mistaken for rubies. For example, the so-called "Black Prince's Ruby" in the British Imperial state crown is actually a 170-carat spinel. Until 1851, the "Kuwait Ruby," also part of the British crown jewels, was considered to be the world's largest ruby, weighing 352.5 carats. It is relatively easy to mistake spinel for ruby because the two gems share many of the same desirable properties: they are chemically similar, deriving their red hue from chromium, and they have similar luster, density, and hardness. Like rubies, spinels seem to fluoresce or glow in natural daylight. Spinels tend, however, to have a slightly pinker hue than rubies.

The derivation of the name is uncertain. It may come from the Latin *spina* for "spine" or "thorn," because the stone is often found as sharp crystals. Since medieval times, it has also been known as the "Balas

▲ *Fancy-cut red spinel.*

▲ *Scissors-cut blue spinel.*

▶ *Faceted spinels from Burma (Myanmar) showing a range of colors.*

SPINEL	
CHEMICAL COMPOSITION	$MgAl_2O_4$
COLOR	RED, BROWN OR BLACK, SOMETIMES GREEN OR BLUE
REFRACTIVE INDEX	1.715–1.720
RELATIVE DENSITY	3.6
HARDNESS	8
CRYSTAL GROUP	CUBIC
CLEAVAGE	NONE
FRACTURE	CONCHOIDAL
TENACITY	BRITTLE
LUSTER	VITREOUS
TRANSPARENCY	TRANSPARENT
DISPERSION	MEDIUM, .020
BIREFRINGENCE	NONE
PLEOCHROISM	NONE

LOCATION Afghanistan, Australia, Burma (Myanmar), Italy, Madagascar, Pakistan, Russia, Sri Lanka, Sweden, Turkey, United States.

▲ *Aggregate of spinel crystals.*

Brilliant

Cushion

Step

Cabochon

Mixed

▲ *Spinel ring with diamond openwork.*

ruby," after Balascia (today Badakhshan), a region of northeast Afghanistan that for many years was a source of fine specimens of spinel.

Although it is red spinels that are the most renowned, the stone occurs in a variety of colors from hot reds and pinks through to cooler blues and greens and even a black, non-translucent variety called **ceylonite**. The reddest stones are known as ruby spinels, orange-reds are known as **rubicelles**, and green are sometimes called **chlorospinels**.

Rubies may be more prized – and more expensive – but good, gem-quality spinels are more rare. The best spinels are chosen for their lively color. They are cut into brilliants, cushions, step cuts, and are occasionally fashioned as round cabochons. Spinels can also be synthesized and synthetic spinels are used to imitate other gems. The magnetic properties of the spinel mineral magnetite, also known as lodestone, was used to help ancient mariners with navigation.

◀ *The large red spinel known as "The Black Prince's Ruby" set in the Imperial State Crown. A small ruby is set in the spinel itself.*

SPINEL

Spinel occurs as an accessory mineral in basic igneous rocks. Spinel's hardness and resistance to weathering means that it can be found as rolled pebbles in river and beach sands. It is often found in association with rubies and sapphires. Gem-quality spinels occur in limestone or limestone gravel and are most commonly found in Burma (Myanmar), Sri Lanka, India, and Russia. Undoubtedly, the finest spinel comes from the Mogok mines of Mandalay, central Burma (Myanmar).

HEMATITE

Bead

Cabochon

Cameo

Brilliant

Table

HEMATITE

Hematite is widely distributed, occurring as an accessory mineral in igneous and sedimentary rocks, and in hydrothermal veins. It is called "iron rose" when arranged like flower petals. Specular hematite, which has dark gray or black crystals with a metallic luster, is also called "looking-glass ore" and was once used as mirrors by the Aztecs. A "kidney ore," especially sought by collectors, is found in metallic black or red-brown lumpy masses resembling kidneys.

Large quantities of hematite are found in the Canadian provinces of Québec, Nova Scotia, and Newfoundland. Hematite is also present in several US states, including the Lake Superior regions of Michigan, Wisconsin, and Minnesota. Crystals measuring 6 inches (15 centimeters) have been discovered in Minas Gerais, Brazil. "Kidney ore" is found in the Forest of Dean and Barrow-in-Furness, England, and "iron roses" at St Gotthard, Switzerland, and Elba.

The most important iron ore, hematite (or haematite) takes its name from the Greek word *haem* for "blood," which describes the color of its powdered form. Ground and powdered, it was used by prehistoric man to make cave paintings, by the Egyptians to decorate tombs of pharaohs, and by Native Americans as a war paint. Hematite was once believed to stop bleeding wounds. Today, it is still used as red ocher pigment and as a metal polishing powder called "jeweler's rouge."

Hematite is popular for jewelry but can be heavy. Types of cut include cabochons, table cuts, and brilliants. It is also carved for such items as figurines, pendants, necklaces, cameos, and intaglios in signet rings. The best cuttable sources come from Cumbria, northwest England; Rhineland-Palatinate, Germany; and the island of Elba, Italy. Black varieties were once used for mourning jewelry.

HEMATITE	
CHEMICAL COMPOSITION	Fe_2O_3
COLOR	STEEL-GRAY, RED, BLACK
REFRACTIVE INDEX	2.94–3.22
RELATIVE DENSITY	4.9–5.3
HARDNESS	5–6.5
CRYSTAL GROUP	TRIGONAL
CLEAVAGE	NONE
FRACTURE	UNEVEN, BRITTLE
TENACITY	BRITTLE
LUSTER	METALLIC, SOMETIMES DULL
TRANSPARENCY	OPAQUE
DISPERSION	——
BIREFRINGENCE	——
PLEOCHROISM	——

LOCATION Ascension Island, Australia, Austria, Brazil, Canada, France, Germany, Italy, Mexico, Norway, Romania, Russia, Switzerland, United Kingdom, United States, Venezuela.

▲ *Polished hematite cabochons.*

◄ *Hematite crystals on matrix.*

CHRYSOBERYL

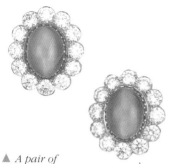

▲ *A pair of chrysoberyl cat's-eye and diamond cluster earrings.*

▼ *The chrysoberyl group shown here (left) with a collection of spinels (right). The chrysoberyl group includes a cat's-eye chrysoberyl gem (bottom left), a chrysoberyl crystal on matrix, and a cross pendant set with cat's-eyes.*

Chrysoberyl has a variable chemical composition and the chrysoberyl group contains a range of gem-quality minerals. All types of chrysoberyl are of great hardness, exceeded in this respect only by DIAMOND, RUBIES, and SAPPHIRES among the gem minerals. This makes chrysoberyl hard-wearing and particularly valuable for jewelry. Chrysoberyl also possesses a relatively high refractive index, giving its gemstones a desirable brilliance.

Chrysoberyl is found, like BERYL, in chromium- and beryllium-rich rocks, such as granite, pegmatites, and mica schists. Most chrysoberyl is recovered from alluvial-rich river sands and gravels.

Mineral collectors especially prize chrysoberyl for its mode of twinning – its crystals habitually form attractive heart-shaped contact twins. The process can be repeated cyclically to yield three pairs of twins, forming a sixfold structure called a "trilling."

The gemmy varieties of chrysoberyl are among the world's rarest gems. Chrysoberyl itself is usually transparent and simply faceted. Its colors range from yellowish green to green, yellow and shades of brown. It is a fine gemstone, but is eclipsed by its more valuable cousins, ALEXANDRITE, named after Tsar Alexander II (on whose birthday it was first discovered), and **cymophane** (cat's-eye chrysoberyl), with its highly prized "milk and honey" coloring. High-quality alexandrite displays the most extreme example of color change (pleochroism) in the chrysoberyl group, and can display different colors under natural and artificial light sources. Cymophane shows strong chatoyancy or "cat's-eye" effect.

Cabochon

Brilliant

Cushion

Mixed

Because chrysoberyl is often yellowish in color, the Greek word *chryso*, meaning "golden," was added to its name to distinguish it from BERYL. Chrysoberyl is very rarely found as gem quality, making it a rare and costly gemstone.

Chrysoberyl is sometimes faceted. But the important variety **cymophane** is usually cut *en cabochon* to take advantage of its remarkable optical properties. "Cymophane" means "wavy appearance" – a reference to the mineral's shimmering cat's-eye effect. The **cat's-eye** (a narrow line of light visible in the polished stone) is due to needle-like inclusions. The term "cat's-eye," applied by itself, properly refers only to chrysoberyl cat's-eye; other stones showing chatoyancy should be called "tourmaline cat's-eye," "ruby cat's-eye," and so on. The effect of chatoyancy is probably seen most clearly in cymophane. The most valuable examples of cymophane or cat's-eye tend to show a thinner line of light.

CHRYSOBERYL

Cat's-eye occurs in granitic rocks and pegmatites, and in mica schists. It is also frequently found in alluvial sands and gravels. It has a white streak. Mineral collectors distinguish chrysoberyl from beryl by the fact that its crystals are tabular, rather than prismatic. It may be confused with PERIDOT.

The most important locations in which chrysoberyl is found include the Ural Mountains of Russia, southern India, Sri Lanka, Burma (Myanmar), Zimbabwe, and Brazil. Although chrysoberyl is found in the United States, there are no major sources of the mineral there.

▶ *Twinned crystal of chrysoberyl, forming "trilling."*

CHRYSOBERYL	
CHEMICAL COMPOSITION	BeAl$_2$O$_4$
COLOR	GREEN, YELLOW
REFRACTIVE INDEX	1.74–1.75
RELATIVE DENSITY	3.5–3.8
HARDNESS	8.0–8.50
CRYSTAL GROUP	ORTHORHOMBIC
CLEAVAGE	PRISMATIC, POOR
FRACTURE	UNEVEN TO CONCHOIDAL
TENACITY	BRITTLE
LUSTER	VITREOUS
TRANSPARENCY	TRANSPARENT TO TRANSLUCENT
DISPERSION	.015
BIREFRINGENCE	.01
PLEOCHROISM	STRONG IN DEEP COLORS

LOCATION Brazil, Burma (Myanmar), India, Mozambique, Russia, Sri Lanka, USA, Zimbabwe.

▶ *Chrysoberyl cat's-eye.*

▼ *Yellow chrysoberyl in a variety of cuts (cushion, right; mixed, left; marquise, bottom).*

Alexandrite, a rare variety of chrysoberyl, is prized for its remarkable optical properties. In daylight, which is rich in shorter wavelengths, it looks bright green. In the warmer, longer-wavelength light of candles or tungsten bulbs, it takes on a rich red or brownish-red color. It has therefore been described as "emerald by day, ruby by night." The color change is due to chromic oxide, which in alexandrite partially replaces the aluminum oxide that occurs in the chemical composition of chrysoberyl. Color changes occur in some types of SAPPHIRE, TOURMALINE, APATITE, and many other gemstones, but alexandrite shows the most dramatic effect.

An important source of alexandrite was the Ural Mountains of Russia, where the variety was first described in 1830. The story has it that it was discovered by emerald miners on the birthday of the future Tsar Alexander II, and was named in his honor by the mineralogist Nils Nordenskjold. The name was doubly appropriate since red and green were the Russian imperial colors. The gemstone became popular in Russian jewelry and the Urals mines were soon exhausted. In the 1920s, Tiffany produced some beautiful rings with alexandrite set in platinum.

Brilliant

Cushion

Mixed

▲ *Alexandrite gems are strongly pleochroic.*

ALEXANDRITE

For much of the 20th century, with no new major discoveries of the mineral, alexandrite was extremely rare. Then in 1987, a new find was made at Hematita, Minas Gerais, Brazil. Alexandrite from Hematita has a dramatic color change from raspberry red to bluish green. In 1993, there was another major find of alexandrite on the border of Tanzania and Mozambique.

▲ *Oval brilliant-cut alexandrite.*

▲ *Twinned crystals of alexandrite, forming "trilling."*

ALEXANDRITE	
CHEMICAL COMPOSITION	$BeAl_2O_4$
COLOR	GREEN IN DAYLIGHT, REDDISH
	IN INCANDESCENT LIGHT
REFRACTIVE INDEX	1.745–1.757
RELATIVE DENSITY	3.71
HARDNESS	8.5
CRYSTAL GROUP	ORTHORHOMBIC
CLEAVAGE	DISTINCT
FRACTURE	CONCHOIDAL, BRITTLE
LUSTER	VITREOUS TO GREASY
TRANSPARENCY	TRANSPARENT
DISPERSION	.015
BIREFRINGENCE	.009
PLEOCHROISM	STRONG
LOCATION Brazil, Mozambique, Russia, Tanzania.	

CORUNDUM

Corundum is an aluminum oxide found worldwide. Only diamond is harder and more durable than corundum. Ruby, sapphire, and emery are varieties of corundum. The red gem variety is known as ruby and all the other colors of gem-quality corundum are known as sapphire. PADPARADSCHA is a rare, pinkish-orange sapphire found only in Sri Lanka. The name "corundum" comes from several ancient names for the stone, such as *kurundam* in Tamil and *kurund* in Hindi. The ancient Greeks mined corundum on the island of Naxos in the Aegean Sea, and Naxos remains the major source of the abrasive emery, used in powdered form on nail files.

Pure corundum is colorless. Today, it is used primarily as an ornamental stone, for bearings in watches, and as an abrasive – either as fragments produced by grinding massive corundum or as the impure form of emery, which is a mixture of corundum, magnetite, and HEMATITE.

Corundum occurs in certain nepheline syenites and nepheline syenite pegmatites. It is also found in metamorphic rocks such as marble, gneiss, and schist. Large crystals occur in some pegmatites, and emery deposits exist in some regionally metamorphosed rocks. Corundum often occurs in rivers. It is found in the United States, Canada, Scotland, Greece, Russia, Australia, Switzerland, India, Madagascar, South Africa, Afghanistan, Burma (Myanmar), Thailand, Sri Lanka, Cambodia, and Japan.

▼ *Ruby and sapphire cut stones. The various colors of corundum are due to trace metallic impurities.*

Any corundum gem that is not red is called sapphire, although popularly sapphire is associated only with blue stones. The usual range of sapphire colors is from very pale blue to deep indigo. Other colors are generally called "fancy sapphires," and these can be black, purple, violet, green, dark gray, yellow, orange, and white. Fine gems are most available under 2 carats but can be found from 5 to 10 carats. Sapphire has been greatly prized as a gemstone since around 800 BC. Rulers of ancient Persia believed the sky had been painted blue by reflection from sapphire stones. Others wore it, like a ruby, as a talisman to ward off illness or as protection while traveling. People in the Middle Ages thought wearing it suppressed wicked thoughts.

Sapphire's various shades are due to iron and titanium impurities. The colors are often in bands, zones, or blotches. The gem's many inclusions reflect light that yields a faint sheen known as "silk." **Leucosapphire** is the name given to the most transparent colorless variety.

Sapphire normally occurs as crystals that have tabular, pyramidal, rhombohedral, or barrel shapes. Repeated twinning usually exists. The gem is normally associated with such gem minerals as RUBY, TOPAZ, ZIRCON, GARNET, SPINEL, and TOURMALINE. Some gems, called **"color-change sapphire,"** will produce different shades of blue in natural and artificial light. These are not as well known but have been popular when set in gold jewelry. Most of these stones come from Sri Lanka or Africa. All sapphires are pleochroic, changing hue when the stone is turned. **Alexandrite sapphire** seems blue in daylight but reddish or violet under an artificial light; ALEXANDRITE, a variety of CHRYSOBERYL, also changes color in different light.

Sapphires come in a variety of shapes and sizes. They have a vitreous luster, and the most valuable (costing tens of thousands of dollars) have a clear, deep blue color. White sapphire is bought as a substitute for the more brilliant diamond and is popular in engagement rings, such as the one Prince Charles gave to Lady Diana. Black sapphire,

▲ *Sapphire and diamond brooch. The calibré-cut sapphires surround the diamond pistil.*

▲ *Cushion-cut sapphire ring, 9.60 carats, flanked by two tsavorite garnets.*

▲ *Spindle-shaped corundum crystal.*

▶ *Treated blue sapphire, brilliant-cut.*

SAPPHIRE	
CHEMICAL COMPOSITION	Al_2O_3
COLOR	BLUE, YELLOW, PINK, GREEN
REFRACTIVE INDEX	1.76-1.77
RELATIVE DENSITY	3.9-4.1
HARDNESS	9
CRYSTAL GROUP	TRIGONAL
CLEAVAGE	NONE
FRACTURE	UNEVEN TO CONCHOIDAL
TENACITY	BRITTLE
LUSTER	VITREOUS
TRANSPARENCY	TRANSPARENT TO TRANSLUCENT
DISPERSION	WEAK, .018
BIREFRINGENCE	.008
PLEOCHROISM	STRONG

LOCATION Australia, Burma (Myanmar), Cambodia, China, India, Madagascar, Nigeria, Sri Lanka, Tajikistan, Thailand, United States.

by contrast, is inexpensive. Star stones occur when several tiny needle-like inclusions of RUTILE reflect light (asterism), usually showing a glittering figure with six points. This effect can be seen when the stone is cut as a cabochon. The American Museum of Natural History, New York, is the home of the "Star of India," a star sapphire weighing 563 carats.

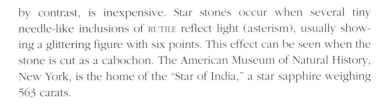

SAPPHIRE

The most prized sapphires come from Kashmir. Kashmiri stones have a rich blue hue and unique "velvet" luster, caused by the presence of minute inclusions. In the 1880s, production peaked in Kashmir and the mines now yield little. Gems from the Mogok Valley of Burma (Myanmar) are well regarded, and Mogok yields many star sapphires. Sri Lanka is the world's largest producer of sapphires larger than 100 carats. Sri Lankan gems tend to be a lighter, cornflower blue. Rakwana stones are particularly prized. Dark and inky blue sapphires are found in New South Wales and Queensland, Australia. Gems from Pailin, western Cambodia, are also highly regarded, but small. China and Nigeria produce dark, iron-rich sapphires. The Bo Ploi mines in Kanchanaburi, Thailand, produce fine stones. In the United States, Yogo Gulch, Montana, produces small sapphires of a metallic blue color. Sri Lanka is the best source for fancy (non-blue) sapphires, and deposits of blue sapphire, pink sapphire and ruby have been discovered in Madagascar. The Umba Valley, Tanzania, and Montana, USA, are also notable locations.

◄ *Faceted sapphires showing a range of colors.*

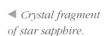

◄ *Crystal fragment of star sapphire.*

▼ *The cabochon "Star of India," a star sapphire weighing 563 carats.*

▲ *Oval light-blue star sapphire brooch with openwork old-cut diamond surround, mounted in silver and gold, c. 1880.*

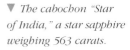

Sapphire is normally faceted as mixed-cut gems, with colorless varieties mostly cut in the brilliant style. Star sapphires and other non-transparent varieties are generally cut as cabochons. If a stone is partly blue and also colorless, a lapidary will put the clear part at the front so the blue at the back makes the entire stone appear blue. Due to the hardness of gem-quality sapphires, carved and engraved sapphires are extremely rare, with most examples dating back to ancient times when lapidaries had royal patrons for customers. This attribute of hardness makes sapphire a fine choice for jewelry that must stand up to everyday wear and tear, such as rings and bracelets.

Most of the sapphires sold today are heat-treated to eliminate impurities and to improve the gem's color and clarity. This form of enhancement is usually permanent. Another stable treatment involves using natural chemicals to alter colors, but this only works on a thin surface layer. If the gem is chipped, the unaltered interior of the stone may be seen. Some fancy sapphire is also irradiated to create intense shades of orange or yellow.

Synthetic sapphire stones, which have been made commercially since 1902, are created by the flame-fusion process using aluminum and adding titanium for color. Imitation stones include a garnet-topped doublet with blue glass base. Star sapphire is imitated by a quartz cabochon with reflective colored pieces at the base.

Brilliant

Cushion

Pendeloque

Cabochon

▲ *Brilliant-cut round sapphire.*

◀ *Ring and earrings with pink and yellow circular-cut sapphire petals, circular-cut garnet leaves, with diamond center.*

Mixed

Padparadscha is a very rare variety of sapphire with a delicate pinkish-orange color, caused by tiny amounts of chromium, iron, and vanadium. The name (sometimes also spelled as "padparadsha") comes from the Sinhalese words *padma radschen*, meaning "lotus-blossom color." Although it is difficult to find gems more than 2 carats in weight, a specimen from Sri Lanka weighing 100 carats is on display at the American Museum of Natural History, New York.

Padparadscha is one of the world's most expensive gems, with prices similar to those fetched by fine ruby or emerald. The most pricey stones display uniformity and purity of color. Unlike other sapphires, the finest color of padparadscha is not a function of intensity (saturation). The most valuable padparadschas display a delicate mixture of pink and orange. Today, many stones are enhanced by heat-treatment. Oval and cushion are the most common cuts.

PADPARADSCHA

CHEMICAL COMPOSITION	AL_2O_3
COLOR	PINKISH-ORANGE
REFRACTIVE INDEX	1.76–1.77
RELATIVE DENSITY	3.9–4.1
HARDNESS	9
CRYSTAL GROUP	HEXAGONAL (TRIGONAL)
CLEAVAGE	NONE
FRACTURE	UNEVEN TO CONCHOIDAL
TENACITY	BRITTLE
LUSTER	ADAMANTINE TO VITREOUS
TRANSPARENCY	TRANSPARENT TO TRANSLUCENT
DISPERSION	WEAK, .018
BIREFRINGENCE	.008
PLEOCHROISM	WEAKLY DICHROIC
LOCATION	Madagascar, Sri Lanka, Vietnam.

◄ *Round and oval brilliant-cut padparadscha gems.*

► *Padparadscha crystal.*

PADPARADSCHA

Purists argue that Sri Lanka is the only legitimate source of padparadscha. However, pinkish-orange sapphires have also been unearthed at the Quy Chau mines in Vietnam, the Tunduru district of the Umba Valley in northeast Tanzania, and on the island of Madagascar. The Tanzanian gems tend to be more brownish orange and are sometimes referred to as "African padparadscha."

▲ *Largest known gemstone of padparadscha, 100.18-carat oval, on display in the American Museum of Natural History, New York.*

▲ *Queen Consort's Ring with a 50.15-carat ruby at its center.*

◀ *Oval ruby and diamond drop earrings.*

Rubies are very rare gems. They exist in many shades of red, from pinkish to purplish or brownish red. The red color is determined by the amount of chromium, which further enhances the color by causing a red fluorescence. A ruby's brownish color indicates iron. The name comes from the Latin word for red, *rubeus*. The true ruby has sometimes been called an "oriental ruby" to distinguish it from the red spinel. Garnets and purple sapphires have also been mistaken for rubies. Besides their obvious value as precious gems, rubies are also used extensively in laser technology.

The great value of rubies is mentioned in the Bible. Ruby mining goes back more than 2,500 years in Sri Lanka and was recorded in the 6th century AD at Mogok, Burma (Myanmar). The Burmese wore the stone as a talisman to protect them against illness or misfortune. Many in the ancient world believed they could predict the future based on the changing color of a ruby they wore. The rare and valued "pigeon's blood" variety now mined in Burma was previously called "blood drops from the heart of Mother Earth" by the Burmese. The Hindus, who thought a ruby burned with an internal fire, called it "the king of precious stones." Rubies have also been assigned various mystical powers and are said to bring the wearer romance, friendship, energy, courage, and peace.

▲ *Ring with a star ruby cabochon weighing 52.27 carats.*

▲ *Ruby doublet.*

▶ *Ruby crystal from Burma (Myanmar).*

RUBY	
CHEMICAL COMPOSITION	Al_2O_3
COLOR	RED TO BROWNISH RED
REFRACTIVE INDEX	1.76–1.77
RELATIVE DENSITY	3.9–4.1
HARDNESS	9
CRYSTAL GROUP	TRIGONAL
CLEAVAGE	NONE
FRACTURE	UNEVEN TO CONCHOIDAL
TENACITY	BRITTLE
LUSTER	VITREOUS
TRANSPARENCY	TRANSPARENT TO TRANSLUCENT
DISPERSION	WEAK, .018
BIREFRINGENCE	.008
PLEOCHROISM	STRONG

LOCATION Afghanistan, Australia, Burma (Myanmar), Cambodia, India, Kenya, Madagascar, Pakistan, Sri Lanka, Tajikistan, Tanzania, Thailand, Vietnam.

RUBY

The center of the ruby trade is located in Bangkok, Thailand, which was once one of the most important ruby mining localities. The Thais developed many ruby cutting and polishing techniques. Thai and Cambodian rubies (largely from around Pailin, western Cambodia) are of high clarity but lack any light-scattering silk inclusions, and only those facets where light is totally internally reflected are a rich red. The best-quality rubies come from the Mogok Valley of Burma (Myanmar) and Mong Hsu in the northeast. Mogok rubies are prized for red fluorescence and tiny amounts of light-scattering rutile silk, which give the stones a beautiful crimson glow. The Vietnamese mines of Luc Yen and Quy Chau also yield superb stones. Sri Lanka is one of the best sources of alluvial "star" rubies. Kenya and Tanzania yield excellent fluorescent stones, but facet-grade material is rare. The situation is similar in Jegdalek, Afghanistan. The state of Karnataka, southwest India, has a history of ruby mining but produces largely low-grade star rubies. The state of Orissa, east India, is a new source. Other sites include Harts Range, Australia; Madagascar, where new fields are being worked; and Tajikistan, which is making an impact as a ruby locality.

Rubies occur in bands of crystalline limestone, and the associated minerals include mica, feldspar, spinel, garnet, graphite, pyrrhotite, and wollastonite. Ruby is a hard stone but can fracture where the crystals are twinned. The hexagonal, barrel-shaped crystal prisms have flat or tapering ends. Short, prismatic crystals have been found in Tanzania in green zoisite, and rubies measuring more than 2 inches (5 centimeters) were discovered in mica schists in the Hunza Valley, Pakistan. Like sapphires, rubies are pleochroic and their hue will change if the stone is turned.

Large rubies have earned higher prices per carat at auctions than flawless diamonds. Fine stones that are more than 2 carats are rare and expensive. The prized colors are pure reds without traces of

▶ *"Lion mask" bracelet by Bulgari. The band of the bracelet consists of cabochon-cut rubies enhanced by circular-cut diamond spacers. The clasp features a cabochon-cut emerald.*

▼ *Madagascar cushion-cut rubies in a variety of shapes.*

brown or blue. Less value is placed on light or dark shades. Intensely colored clear pink varieties are also greatly valued, especially when cut well. Rubies are usually fashioned as a mixed-cut or cabochon. They are very durable, and are used for rings, bracelets, necklaces, earrings, and other items of jewelry. Breakage rarely occurs because there is no easy cleavage.

Rubies should be evaluated under different intensities of light. A very strong light normally gives a ruby a very intense color, while normal light may show a less intense coloring. The stones should also be assessed for their symmetry by placing them face up.

Like sapphires, rubies may have needle-like inclusions that appear as a "silk" sheen. A cabochon-cut allows these inclusions to resemble a six-pointed star. The "Eminent Star" ruby, an oval cabochon with a six-ray star, weighs 6,465 carats. Attractive star inclusions can add to a ruby's value. Some sellers may fill fissures with materials such as glass or borax to improve the gem's appearance and durability, but the filler material may break or fall out with rough treatment or exposure to heat.

Synthetic rubies were the first synthetic gemstones to become commercially available in large quantities. Small fragments of the real stone were once fused together to create "reconstructed rubies," but this practice came to an end in 1902 when French chemist Auguste Verneuil flame-fused powdered aluminum oxide and a coloring material to produce a synthetic stone. Today, synthetic rubies can be very difficult to distinguish from true rubies.

Brilliant

Step

Mixed

Cabochon

Cameo

◀ *Edwardes ruby in the Natural History Museum, London; weight 167 carats.*

▼ *Ruby cut as a cabochon.*

CASSITERITE

Brilliant

Mixed

The major ore of tin, cassiterite is also appreciated for its sparkle, which derives from its high luster. It is often found as an eight-sided prism, consisting of two four-sided prisms, one of which is usually dominant. While it is generally black, reddish-brown crystals, large enough for cutting, are also occasionally found. It is named from the Greek word *kassiteros* for tin.

Cassiterite's importance as a tin ore has been known for millennia. During the Bronze Age, it was added to molten copper to form bronze, a copper alloy that is strong and durable yet easily worked. Some of the oldest cassiterite mines, such as those in Cornwall, England, have been worked since 2000 BC and are now exhausted of minerals. Cassiterite is generally located within or near granite masses. It is associated with wolframite, arsenopyrite, bismuthinite, TOPAZ, QUARTZ, TOURMALINE, and mica. Today, because so many established mines have been intensively worked over the centuries, most cassiterite is found in alluvial deposits.

CASSITERITE

In addition to its industrial use as a source of tin, cassiterite is also used as a gemstone. It needs polishing to bring out its best, which may be time-consuming, but cutting is straightforward because it is hard and cleavage is imperfect. Large cut stones, of more than 1 carat, are rare. The best stones for cutting come from Erongo, central Namibia, and from Galicia, Spain.

▲ *Cassiterite faceted as an oval brilliant-cut.*

◀ *Cassiterite crystals.*

▲ *Agglomeration of tetragonal crystals of cassiterite.*

CASSITERITE	
CHEMICAL COMPOSITION	SnO_2
COLOR	USUALLY BLACK
REFRACTIVE INDEX	1.99–2.09
RELATIVE DENSITY	6.8–7
HARDNESS	6–7
CRYSTAL GROUP	TETRAGONAL
CLEAVAGE	PRISMATIC, IMPERFECT
FRACTURE	UNEVEN
TENACITY	——
LUSTER	ADAMANTINE
TRANSPARENCY	OPAQUE
DISPERSION	.071
BIREFRINGENCE	STRONG, .096
PLEOCHROISM	DICHROIC, DEFINITE
LOCATION Bolivia, China, Indonesia, Malaysia, Mexico, Namibia, Russia, Spain.	

Although its high refractive index endows it with sparkle that is greater than DIAMOND, rutile's fire is dimmed by its reddish-brown to black color. It is this color that gives it its name, which derives from the Latin *rutilus*, meaning "reddish."

Rutile occurs as an accessory mineral in a variety of igneous rocks, as well as in schists, gneisses, metamorphosed limestones and quartzites. It is also a secondary mineral produced by the breakdown of titanium-bearing minerals such as titanite and some micas, and may be concentrated in alluvial deposits and beach sands. In addition to being a gemstone, rutile is a major ore of titanium, which is used in high-tech alloys and as a pigment in paint, porcelain, and false teeth.

In jewelry, rutile has been used for thousands of years. The black or dark brown stones were traditionally used in mourning jewelry. The stones are usually polished to show off their dark color. Because they have fine internal cracks along their prismatic faces, they are difficult to cut. Their low translucence means that cutting may not be that beneficial to their appearance, but rutiles are sometimes found shaped into baguettes. The gem is said to encourage feelings of tranquillity and forgiveness, and to help strengthen the immune system.

While rutile on its own is an attractive stone, it is perhaps more interesting when it appears with other gems. When rutile occurs as a microscopic **inclusion** in gemstones such as TOURMALINE, RUBY or SAPPHIRE, it creates striking light effects like cat's-eyes or asterisms. RUTILATED QUARTZ is found when golden rutile needles are embedded in clear quartz, giving an appearance of golden hair in the crystal. This popular gemstone is sometimes known as "Venus' hair" or "Cupid's darts."

Baguette

Mixed

RUTILE	
CHEMICAL COMPOSITION	TiO$_2$
COLOR	REDDISH-BROWN; ALSO CAN BE YELLOW OR BLACK
REFRACTIVE INDEX	2.63
RELATIVE DENSITY	4.25
HARDNESS	6–6.5
CRYSTAL GROUP	TETRAGONAL
CLEAVAGE	PRISMATIC, DISTINCT
FRACTURE	CONCHOIDAL TO UNEVEN
TENACITY	————
LUSTER	ADAMANTINE TO SUBMETALLIC
TRANSPARENCY	TRANSPARENT WHEN THIN, OTHERWISE SUBTRANSLUCENT TO OPAQUE
DISPERSION	OUTSTANDING
BIREFRINGENCE	.285–.296
PLEOCHROISM	WEAK TO MODERATE

LOCATION Brazil, Switzerland, United States.

RUTILE

Some of the best rutile samples have been excavated from the mines in Minas Gerais, southeast Brazil. There, yellowish brown, yellowish green and transparent stones have been found, which are prized by collectors. Crystal twinning also forms some interesting, symmetrical samples.

▲ *Polished crystal fragment of quartz with rutile inclusions.*

◄ *Rutile crystal fragment on matrix.*

FLUORITE

Step

Cushion

Mixed

Cameo

Fluorite is a widely distributed mineral that is mined in great quantities. In occurs in mineral veins, either alone or as a gangue mineral with metallic ores and in association with QUARTZ, BARYTE, CALCITE, CELESTINE, DOLOMITE, galena, CASSITERITE, SPHALERITE, TOPAZ, and many other minerals. Its crystals are commonly cubic, often with rounded corners. It is less frequently octahedral or rhombdodecahedral, and interpenetrant twins are common. The name comes from the Latin word *fluere* meaning "to flow" – a reference to its low melting point and use as a flux in the smelting of metals. Fluorite has given its name to the phenomenon of fluorescence.

Fluorite is worked mainly for use as a flux in the smelting of iron and in the chemical industry. It is used in the preparation of fluorine compounds like hydrofluoric acid. Smaller amounts are used as decorative stones and in the manufacture of specialized optical equipment, pottery, enamels, plastics, toothpaste, refrigerants, and coatings for non-stick pans.

Although fluorite is too soft and too readily cleavable to be used as a faceted gemstone, some stones have been faceted as trap cuts, brilliants, and occasionally as cabochons. Irregular pieces are tumbled for necklaces. The color variation, particularly in the color-banded variety known as **blue john**, has made it prized as an ornamental stone from which vases and ornaments have been fashioned since ancient times. English miners once called the crystals "ore flowers," and collectors have said fluorite is "the most colorful mineral in the world."

Blue john is a variety of fluorite with curved bands of blue, purple, violet, yellow, and white. Some of the darker colors may appear black. The distinctive colors of this fluorite were made millions of years ago either by the inclusion of manganese or oil. There are at least 14 different banding patterns. Blue john is fragile and may be

FLUORITE	
CHEMICAL COMPOSITION	CaF_2
COLOR	YELLOW, GREEN, BLUE, PURPLE, PINK, RED, BLACK, COLORLESS
REFRACTIVE INDEX	1.43
RELATIVE DENSITY	3.2
HARDNESS	4
CRYSTAL GROUP	CUBIC
CLEAVAGE	PERFECT
FRACTURE	SUBCONCHOIDAL
TENACITY	BRITTLE
LUSTER	VITREOUS
TRANSPARENCY	TRANSPARENT TO TRANSLUCENT
DISPERSION	.007
BIREFRINGENCE	NONE
PLEOCHROISM	NONE

LOCATION Austria, Bulgaria, Canada, China, Czech Republic, England, France, Germany, Italy, Kazakhstan, Norway, Pakistan, Poland, Russia, Switzerland, United States.

▲ *Cubic crystals of fluorite.*

FLUORITE

Besides Derbyshire's stone, excellent crystals are found at Weardale near Durham, and at Alston and Cleator Moor in Cumbria, England. Fluorite is found worldwide. Major sources in North America include Haliburton County and Hastings County, Ontario, Canada; Trigo Mountains, La Paz County, Arizona; San Juan County, Colorado; Cave-in-Rock, Hardin County, Illinois; Franklin, New Jersey; and Socorro County, New Mexico. Sites in Australia include the Flinders Range, South Australia. Other locations include Alsace, northeast France.

▲ *Cabochon-cut fluorite gem.*

bonded with resins to help protect against damage. The name was supposedly derived from the French *bleu et jaune*, meaning "blue and yellow." The only source of blue john is at Castleton in Derbyshire, England, giving the town its nickname of "Gem of the Peaks." The stone, in a fibrous to columnar form, was discovered by early miners seeking lead and is now found in two caverns in the Peak District National Park: Treak Cliff Cavern (where 500 kilograms are mined annually) and Blue John Cavern. The larger veins in the Derbyshire mines have been overworked and no longer exist, so decorative objects made today are smaller.

Blue john was used by the ancient Romans for ornamental objects, with vases discovered in the ruins of Pompeii. It became very popular in the late-18th and 19th centuries, when it was often used to fashion vases, urns, and dishes. Splendid examples of blue john ornaments are on display at Windsor Castle, England, and in many English stately homes, such as Chatsworth House in Derbyshire, as well as in the White House in Washington, D.C., and the Vatican, Rome.

▼ *Blue john vase in the Natural History Museum, London.*

▲ *Polished slice of blue john.*

CALCITE

Step

Polished

One of the most common and widely distributed minerals, calcite is the major constituent of limestone, marble, and chalk. The name is derived from the Latin word *calx*, meaning "lime." Calcite is quarried as limestone to make cement, building blocks, ornamental stones, and fertilizers. Pure metamorphosed limestone forms white granular marble, and the presence of other minerals results in colored, figured marble used for buildings, decorations, and carvings. Banded Mexican onyx is a variety of calcite with a marble-like texture used for ornamental work and carved into figurines.

Calcite crystals are more varied than any other mineral, with the best formed in rock cavities. They are large, transparent, and colorless prismatic crystals intergrown with other minerals. More than 300 forms are known, and they combine into a thousand different variations. Because of its softness, calcite is only faceted for collectors. Pink or red **cobalto-calcite** makes beautiful items. A white, fibrous variety is cut into cabochons to show the cat's-eye effect. **Iceland spar**, a variety first found in lava cavities in Iceland, is colorless and transparent, having large clear prisms. It is used to make special lenses, including those for the dichroscope, an instrument that displays the pleochroic colors of gemstones. Iceland spar is also doubly refracting, meaning the crystal splits light into two rays to produce a double image.

CALCITE

Calcite is normally found in veins within rock. Associated minerals include FLUORITE, QUARTZ, sulfur, GOLD, copper, EMERALD, and APATITE. In limestone caves, it produces stalactites and stalagmites, and it also forms seashells of living organisms which, on death, form limestone.

Calcite occurs worldwide. Eskifjord in Iceland yields Iceland spar. In Canada, major calcite deposits include Cobalt, Timiskaming County, and Dungannon Township, Hastings County, Ontario. In Australia, the Mount Lofty Ranges in South Australia have notable deposits. US locations include Cochise County, Arizona, and Franklin, New Jersey. UK sources include Egremont, West Cumbria, and Leadhills, Strathclyde.

◀ *Yellow step-cut calcite gem.*

▶ *Calcite crystals.*

CALCITE	
CHEMICAL COMPOSITION	CaCO₃
COLOR	COLORLESS OR WHITE; SHADES OF GRAY, YELLOW, GREEN, RED, PURPLE, BLUE, BROWN, BLACK
REFRACTIVE INDEX	1.48–1.66
RELATIVE DENSITY	2.7
HARDNESS	3
CRYSTAL GROUP	TRIGONAL
CLEAVAGE	PERFECT
FRACTURE	CONCHOIDAL BUT RARE
TENACITY	BRITTLE
LUSTER	VITREOUS
TRANSPARENCY	TRANSPARENT TO TRANSLUCENT OR NEARLY OPAQUE
DISPERSION	WEAK, .008–.017
BIREFRINGENCE	.15–.19
PLEOCHROISM	NONE
LOCATION Austria, Canada, China, England, France, Germany, Iceland, Ireland, Mexico, Namibia, United States.	

Rhodochrosite is a manganese carbonate with a vivid pink-rose and red color derived from manganese. Its descriptive name comes from the Greek words *rhodon* ("rose") and *chroma* ("color"). If massive aggregates are mined, the mineral can be an ore of manganese. It is best known, however, in gemstone jewelry. Some fine crystals of rhodochrosite are cut into gems, but this is difficult because of the perfect cleavage. Such items are seldom for everyday usage because of rhodochrosite's brittleness. A few cabochons have been cut from exceptionally clear crystals.

Fine-grained, banded rhodochrosite became a popular ornamental stone during the 1930s, and this is often carved into decorative objects like figurines. Bands can be pink and red, as well as pink and white. Collectors especially value tubular stalactitic forms that are cut into cross-sections to show concentric bands.

Rhodochrosite occurs in hydrothermal mineral veins containing ores of silver, lead, and copper. It has also been noted in metamorphic and metasomatic rocks of sedimentary origin, and in sedimentary deposits of manganese oxide, where it is of secondary origin. Individual crystals are found in rhombohedrons and sometimes scalahedrons, but large crystals are rare. Rhodochrosite is distinguished from RHODONITE, also a pink manganese mineral, by its inferior hardness, and often develops a brown or black crust on exposure to air. Associated minerals include CALCITE, ankerite, alabandite, rhodonite, bementite, FLUORITE, manganite, QUARTZ, galena, pyrite, and chalcopyrite.

Bead

Cabochon

◄ *Rhodochrosite crystals.*

RHODOCHROSITE

The United States is the primary producer of rhodochrosite. The Sweet Home Mine in Alma, Colorado, has excellent gem crystals of a clear quality, including deep-red specimens. Other US states with mines include California, Oregon, and New Jersey. San Luis province in central Argentina has been a key source since the 13th century, when Incas worked the local silver mines. Its banded rhodochrosite is sometimes also called "Inca rose." Another important site is at Capillitas near Andalgalá, northwest Argentina. Mont Saint-Hilaire, Québec, Canada, produces many fine pinks and reds, and North Cape Province, South Africa, is the site of the banded variety and fine transparent crystals. Rhodochrosite is also mined at Pachuca, Hidalgo, southern Mexico; Linópolis, Minas Gerais, southeast Brazil; and the Lima Department of Peru.

RHODOCHROSITE	
CHEMICAL COMPOSITION	$MnCO_3$
COLOR	ROSE-PINK, RED
REFRACTIVE INDEX	1.60–1.80
RELATIVE DENSITY	3.4–3.7
HARDNESS	3.5–4.5
CRYSTAL GROUP	TRIGONAL
CLEAVAGE	PERFECT
FRACTURE	UNEVEN
TENACITY	BRITTLE
LUSTER	VITREOUS
TRANSPARENCY	TRANSLUCENT
DISPERSION	WEAK
BIREFRINGENCE	.218
PLEOCHROISM	NONE

LOCATION Argentina, Australia, Brazil, Canada, Germany, Mexico, Peru, Romania, Russia, South Africa, Sweden, United States.

◄ *Oval mixed-cut rhodochrosite.*

SMITHSONITE

Cabochon

▲ *Sky-blue cabochon gem of smithsonite.*

This intriguing stone is named for James Smithson (1765–1829), the British geologist who first defined the mineral, and who left his fortune upon his death to be used to found the Smithsonian Institution in Washington, D.C.

The main occurrence of smithsonite is in the oxidized zone of ore deposits carrying zinc minerals. It is commonly associated with sphalerite, hemimorphite, galena, and CALCITE. It generally forms in dry climates, as primary sulfite zinc ores are weathered. Although smithsonite's official chemical composition makes it a zinc carbonate, some of the zinc is often replaced by other minerals, which gives the stone a unique color. When basalt is present, the stone takes on a pink to rose tint; the best examples come from the Tsumeb mine in northern Namibia, and the Kabwe (formerly Broken Hill) mine in central Zambia. Cadmium gives the stone a yellow color, and this is known as **cadmium smithsonite**. In the best-known mix, copper makes smithsonite appear green or blue. The most sought-after color variety is purple to lavender.

Collectors are also attracted by the unique rounded and bubbly shape of smithsonite samples. They have a botryoidal crystal shape, lending the stone the appearance of a bunch of grapes, as radiating crystals grow out from a central point and then back into each other. The best samples benefit from a silky luster, which is said to resemble that of molten wax gleaming in the candlelight.

SMITHSONITE

While smithsonite is too soft for jewelry, it can be shaped and polished, and used as ornamental stone. In this context it is sometimes known as **bonamite**. In addition to the mines in Namibia and Zambia, other major deposits of smithsonite include Mausbach, North Rhine-Westphalia, western Germany.

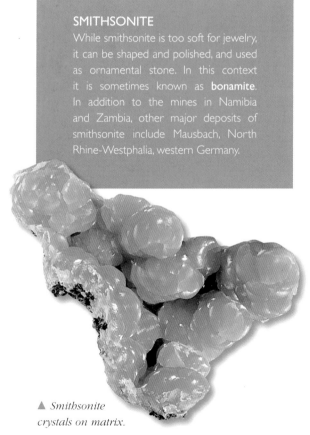

▲ *Smithsonite crystals on matrix.*

SMITHSONITE	
CHEMICAL COMPOSITION	ZnCO₃
COLOR	USUALLY GRAY, BROWN, GRAYISH WHITE
REFRACTIVE INDEX	1.621–1.849
RELATIVE DENSITY	4.4–4.5
HARDNESS	4–5
CRYSTAL GROUP	TRIGONAL
CLEAVAGE	RHOMBOHEDRAL, PERFECT
FRACTURE	UNEVEN
TENACITY	BRITTLE
LUSTER	VITREOUS
TRANSPARENCY	TRANSLUCENT
DISPERSION	.037
BIREFRINGENCE	.228
PLEOCHROISM	NONE
LOCATION Greece, Italy, Mexico, Namibia, Spain, United States, Zambia.	

Dolomite is a mineral composed of calcium and magnesium carbonate, being similar to CALCITE. It was named after D. Dolomieu (1750–1801), a French mineralogist who described it in 1798 during a visit to Egypt with Napoleon Bonaparte. The mineral's original name was *dolomie*. It is used in the manufacture of refractory bricks for furnace linings. **Pearl spar** is a variety with a pearly luster and colored white, gray or pale brown. **Bitter spar** is an iron-bearing variety, colored brown by ankerite.

Dolomite crystals are colorless, white, pink, or yellow, forming rhombohedrons or sometimes having curved faces that combine into distinctive saddle-shaped growths. They are rarely faceted because of dolomite's softness and perfect cleavage, but step-cuts are popular with some collectors.

Dolomite is found in metamorphic marble rock and occurs as a gangue mineral in veins with SPHALERITE or galena. Its most important occurrence, however, is as a rock-forming mineral in carbonate rocks. Sedimentary dolomite rock is called dolomitic limestone, magnesian limestone, dolostone, and, previously, dolomite. Made up almost totally of the mineral dolomite, with small amounts of calcite, it is found worldwide in shades of white, cream, gray, brown, and pink. Dolomitic limestone is used as a building stone and to produce special cements. Also, it is a source of magnesium oxide, and is used as an antacid and in insulating materials.

Brilliant

Step

◀ *Round brilliant-cut dolomite gem.*

DOLOMITE	
CHEMICAL COMPOSITION	CaMg(CO$_3$)$_2$
COLOR	WHITE, COLORLESS, YELLOWISH, BROWN, PINK
REFRACTIVE INDEX	1.50–1.68
RELATIVE DENSITY	2.9–3.2
HARDNESS	3.5–4
CRYSTAL GROUP	TRIGONAL
CLEAVAGE	PERFECT
FRACTURE	SUBCONCHOIDAL
TENACITY	BRITTLE
LUSTER	VITREOUS TO PEARLY
TRANSPARENCY	TRANSPARENT TO TRANSLUCENT
DISPERSION	WEAK
BIREFRINGENCE	.18
PLEOCHROISM	NONE

LOCATION Austria, Brazil, Canada, Czech Republic, England, Germany, Italy, Mexico, Saudi Arabia, Scotland, Spain, Switzerland, United States.

DOLOMITE

Deposits of dolomite are found in Brosso and Traversella, Piedmont, Italy; Binnenthal, Switzerland; the Freiberg and Schneeberg mines in Germany; Cornwall and Cumbria, England; Scotland; St Eustache, Québec, Canada; Guanajuato, Mexico; several US states, including Vermont, New York, New Jersey, North Carolina, Missouri, Iowa, Wyoming, and Colorado; Spain; Austria; Czech Republic; Brazil; and Saudi Arabia.

▲ *Dolomite crystals on matrix.*

ARAGONITE

Bead

Polished

Aragonite is a carbonite mineral chemically the same as calcite, but unlike its more common polymorph it has a fibrous, rounded form and no rhombohedral cleavage. Aragonite is a popular glassy ornamental stone that can also be found in layered or banded varieties. Especially beautiful crystals have been found in Wind Cave, South Dakota, USA, and the Sicilian sulfur deposits of Italy.

▲ *Pendeloque-cut aragonite gem.*

ARAGONITE

Aragonite occurs as a deposit from hot springs and in association with beds of gypsum. It has been noted in veins and cavities with calcite and dolomite, and in the oxidized zone of ore deposits, where it is found together with secondary minerals such as malachite and smithsonite.

Areas rich in aragonite include Aragón, Spain; Bastennes, France; Cumbria, England; Leadhills, Scotland; Agrigento, Sicily; Tsumeb, Namibia; Carinthia, Austria; Harz Mountains, Germany; and in the southwest United States including Chester County, Pennsylvania; Socorro County, New Mexico; and Fort Collins, Colorado.

Aragonite was named in 1790, after the Spanish province of Aragón where it was first noted. The shells of certain molluscs, such as the abalone, are made of aragonite, which gives them a mother-of-pearl luster. Many fossil shells now composed of calcite were formed originally of aragonite, which is unstable in high pressures and temperatures. These fossils have usually lost the original pearly luster. Natural pearls are also mostly formed by aragonite bonding with conchiolin, a process that takes about seven years.

Prismatic twinning is a common feature of the aragonite crystal, and if three twins occur the mineral becomes pseudo-hexagonal. The *flos-ferri* ("iron flowers") variety has thin, intertwined and clumping branches that resemble worms or trees. Some of the well-known sites for *flos-ferri* are in Austria, Russia, Mexico, and the United States.

ARAGONITE	
CHEMICAL COMPOSITION	$CaCO_3$
COLOR	COLORLESS, GRAY, WHITE, YELLOWISH
REFRACTIVE INDEX	1.67–1.69
RELATIVE DENSITY	2.9
HARDNESS	3–4
CRYSTAL GROUP	ORTHORHOMBIC
CLEAVAGE	PINACOIDAL, IMPERFECT
FRACTURE	SUBCONCHOIDAL
TENACITY	BRITTLE
LUSTER	VITREOUS
TRANSPARENCY	TRANSPARENT TO TRANSLUCENT
DISPERSION	NONE
BIREFRINGENCE	.155
PLEOCHROISM	NONE
LOCATION Austria, England, France, Germany, Italy, Mexico, Namibia, Poland, Scotland, Spain, United States.	

▲ *Rough specimen of aragonite.*

A lead ore, cerussite gets its name from the Latin word *cerussa*, meaning "white lead." Although its natural form is rarely used as a pigment, the synthesized form, lead carbonate, has been widely used in paintings since the 5th century BC. Today, cerussite is of great interest to collectors for its fantastic sparkle, which derives from its high lead content, as well as for its complex, twinned crystals.

Brilliant

Cerussite is usually of secondary origin, found in the oxidized zone of lead veins. It is frequently found in association with many other minerals, most commonly ANGLESITE, galena, SMITHSONITE, promorphite, and SPHALERITE. While cerussite itself is clear and transparent, it may derive some color from the minerals with which it is associated: for example, a cerussite sample rich in galena will appear gray, while one with MALACHITE will be green. Clear cerussite minerals are the most valuable.

The twinned crystals of cerussite form very intricate structures. There are three types: chevron shapes are the most common; cyclic crystals often create fantastic, six-pointed stars; while reticulated crystals are the most complex. Some of the best crystal samples, found in Tsumeb, northern Namibia, have been up to 2 feet (60 centimeters) tall.

Cerussite is very soft and brittle, and must be handled with care. Because it is so fragile, it is extremely difficult to cut and, as a result, faceted gems are highly prized by collectors.

CERUSSITE	
CHEMICAL COMPOSITION	$PbCO_3$
COLOR	WHITE, GRAY, GREEN
REFRACTIVE INDEX	2.07
RELATIVE DENSITY	6.58
HARDNESS	3-3.5
CRYSTAL GROUP	ORTHORHOMBIC
CLEAVAGE	PRISMATIC IN TWO DIRECTIONS, DISTINCT
FRACTURE	CONCHOIDAL
TENACITY	BRITTLE
LUSTER	ADAMANTINE
TRANSPARENCY	TRANSPARENT TO TRANSLUCENT
DISPERSION	STRONG, .051
BIREFRINGENCE	VERY STRONG, .274
PLEOCHROISM	NONE

LOCATION France, Germany, Mexico, Morocco, Namibia, Spain, United States.

CERUSSITE

Stones suitable for cutting come from Spain or Namibia. The type locality is Vicentin, Italy. Other sources of cerussite include the Haut Rhin, Alsace, northeast France; Oberpfalz, Bavaria, southern Germany, and Aachen, North Rhine-Westphalia, western Germany; the Sierra de las Encinillas, Chihuahua, north-central Mexico, and Jacala, Hidalgo, southern Mexico; and Mibladen, Khenifra, central Morocco.

► *Oval cushion-cut cerussite gem.*

◄ *Acicular (needle-like) crystals of cerussite on matrix.*

MALACHITE

Cabochon

MALACHITE	
CHEMICAL COMPOSITION	$Cu_2CO_3(OH)_2$
COLOR	BRIGHT GREEN
REFRACTIVE INDEX	1.85
RELATIVE DENSITY	3.9–4.0 (MASSIVE VARIETIES AS LOW AS 3.5)
HARDNESS	3.5–4
CRYSTAL GROUP	MONOCLINIC
CLEAVAGE	PINACOIDAL, PERFECT
FRACTURE	SUBCONCHOIDAL OR UNEVEN
TENACITY	BRITTLE
LUSTER	FIBROUS VARIETIES SILKY, DULL WHEN MASSIVE, CRYSTALS ADAMANTINE
TRANSPARENCY	TRANSLUCENT
DISPERSION	NONE
BIREFRINGENCE	.254
PLEOCHROISM	WEAK

LOCATION Australia, Democratic Republic of Congo, England, France, Germany, Israel, Namibia, Russia, United States.

Malachite is a common secondary copper mineral with a vivid green color. Its name derives from the Greek word *moloche* meaning "mallow," because it has the color of mallow leaves. Malachite was crushed as a green pigment in the Bronze Age and has been valued as a gemstone since ancient times.

The Greeks and Romans fashioned vases and sculptures from malachite, and wore it as amulets to ward off evil spirits and illnesses. Later, the Russian Tsars used it as a decorative stone, quarrying vast deposits in the Ural Mountains to create magnificent structures, such as the columns of St Isaac's Cathedral in St Petersburg, western Russia.

▲ *Earring with malachite drops and bead, c. 1860.*

Malachite is a relatively soft mineral, but some is cut into rounded cabochons and other rounded shapes, while some is fashioned as beads. The quality required for jewelry is mined at several sites, including Eilat in Israel. This can be carved and polished to reveal alternating bands of shades of green, from pale to near-black layers, and can show intricate patterns on the polished surface. The ornamental stone is also used for items like tabletops, bowls, vases, and carved animals.

▲ *Polished slice of malachite.*

▶ *Mass of malachite (green) and azurite (dark blue).*

MALACHITE

Malachite mineral is common in the oxidized zone of copper deposits and normally found in opaque green masses that are granular, knobbly, and massive. It is usually with AZURITE, which has a deep azure blue color. **Chessylite** is the name given to banded malachite and azurite occurring in Chessy, near Lyons, France. Malachite is also associated with copper, CUPRITE, CALCITE, chalcocite, CHRYSOCOLLA, chalcopyrite, and limonite.

The renowned Urals deposits in Russia are becoming depleted. Most large deposits are found in copper-mining areas, with the Shaba province of the Democratic Republic of Congo the major producer. Other sites are at Betzdorf, Rhineland-Palatinate, Germany; Potosi in New South Wales, Australia; Redruth in Cornwall, England; and in the United States at Bisbee in Arizona, Stevens County in Washington, and the states of Pennsylvania, North Carolina, and New Mexico. Other mines also operate in France, Mexico, Zambia, Namibia, and Sweden.

A hydrous copper silicate, chysocolla is a minor copper ore. It is especially found in copper mines in arid areas, such as the southwest United States. Some examples have been used as an alternative to TURQUOISE when it resembles that stone's color. The name first appears (315 BC) in the writings of Theophrastus and comes from the Greek words *chrysos* for "gold" and *kolla* for "glue," because chrysocolla resembles other materials, including borax, used as a flux in soldering gold in ancient times.

▲ Greenish-blue polished chrysocolla.

Chysocolla forms as crusts, stalactites or stalagmites, and in botryoidal (grape-like) shapes, as well as inclusions in other minerals. Crystals do not occur naturally. Chrysocolla is quite soft and fragile and tends to break easily when exposed to the atmosphere. However, intermixture with quartz or CHALCEDONY gives it durability. Lapidaries can polish the stone to accentuate the colored chrysocolla and sparkles of quartz. It is often cut as beautiful greenish-blue cabochons, and used for ornaments such as carving and figurines. In the 1950s, US lapidaries voted chrysocolla-colored chalcedony the "most popular American gem."

Bead

Cabochon

CHRYSOCOLLA

Chrysocolla occurs in copper veins, and is formed by waters containing silica. It is often mixed with copper compounds, and associated minerals include QUARTZ, AZURITE, TURQUOISE, limonite, CUPRITE, tenorite, HEMATITE, and MALACHITE. It is similar to malachite, but is more bluish in color.

A mottled blue-and-green variety comes from Arizona, and chrysocolla is found in other US states including California, New Mexico, New Jersey, Utah, Michigan, and Pennsylvania. Key deposits are also in Wanlockhead, Dumfries and Galloway, Scotland; Chihuahua in Mexico; South Australia, Australia; Alsace, France; and Shaba province, Democratic Republic of Congo (formerly Zaïre).

CHRYSOCOLLA

CHEMICAL COMPOSITION	$(Cu,Al)_2H_2Si_2O_5$ $(OH)_4 \cdot n(H_2O)$
COLOR	SKY-BLUE, GREENISH-BLUE, GREEN
REFRACTIVE INDEX	1.58–1.64
RELATIVE DENSITY	2–2.4
HARDNESS	2–4
CRYSTAL GROUP	MONOCLINIC
CLEAVAGE	NONE
FRACTURE	CONCHOIDAL
TENACITY	BRITTLE
LUSTER	VITREOUS
TRANSPARENCY	TRANSLUCENT TO ALMOST OPAQUE
DISPERSION	——
BIREFRINGENCE	.13–.17
PLEOCHROISM	NONE
LOCATION	Australia, Democratic Republic of Congo, France, Germany, Israel, Mexico, United Kingdom, United States.

◄ Bluish crystals of chrysocolla on matrix.

AZURITE

Cabochon

Cameo

Widely used in the ancient world and through the Middle Ages as a pigment and dye, azurite is easily identified by its spectacular, blue color. Its name derives from the Persian word for blue, *lajward*. A basic carbonate of copper, azurite is found wherever there are copper deposits. It is a secondary mineral, created by weathering in the upper oxidation zone of copper ore. It is formed either by water containing carbon dioxide reacting with copper-bearing minerals, or by cupric salts reacting with limestones. Azurite often forms sharp crystals, which are either tabular, or short and prismatic. It is frequently found in association with MALACHITE, a striking, green mineral that results from further oxidation of azurite and is thus more stable. The mixture of azurite and malachite is known as **azurmalachite**. The striking combination of azurite's deep blue and malachite's vivid green is popular with jewelry enthusiasts and gem collectors. Azurite has been valued for its color for many centuries and was particularly popular among painters in the Middle Ages. Today, it is still used as a color base for some paints. However, its oxidation process is ongoing, and over time its color may tend toward greenness when it is used in paints without fixatives.

AZURITE

Despite its attractive color, azurite is not used for jewelry that frequently because of its low hardness. However, it may be coated with wax or another clear substance to protect it. When shaped, it is generally fashioned into thin table cuts. Correct fashioning will enhance its luster, making it almost adamantine in appearance. Polished azurmalachite, often cut into cabochons, reveals banding of alternating deep blue and bright green.

Azurite is found around the world. Significant deposits are in Tsumeb, Namibia; Utah, New Jersey, and Arizona in the United States; and near Adelaide, Australia.

▶ *Cabochon-cut azurmalachite gem.*

AZURITE	
CHEMICAL COMPOSITION	$Cu_3(CO_3)_2(OH)_2$
COLOR	VARIOUS SHADES OF
	DEEP AZURE-BLUE
REFRACTIVE INDEX	1.7–1.8
RELATIVE DENSITY	3.7–3.9
HARDNESS	3–4
CRYSTAL GROUP	MONOCLINIC
CLEAVAGE	PRISMATIC, PERFECT;
	PINACOIDAL, LESS SO
FRACTURE	CONCHOIDAL
TENACITY	BRITTLE
LUSTER	VITREOUS
TRANSPARENCY	TRANSPARENT TO TRANSLUCENT
DISPERSION	NONE
BIREFRINGENCE	.11
PLEOCHROISM	WEAK
LOCATION Worldwide, especially Australia, France, Mexico, Morocco, Namibia, United States.	

▲ *Azurite crystals on matrix.*

Step

Mixed

Popular with collectors of rare and unusual gems, sinhalite was only identified in 1952. Previously, similar brown gemstones found in Sri Lanka were thought to be a variety of PERIDOT, but chemical and crystallographic analysis revealed the truth. Its name derives from the Sanskrit word for Sri Lanka, *Sinhala*.

Virtually all sinhalite is found in the gem gravel beds of Sri Lanka, associated with other gem gravel minerals such as RUBY, SAPPHIRE, peridot, and GARNET. Large specimens are available, and gems of more than 100 carats have been found. However, crystals are very rare, although some are found in Burma (Myanmar).

▲ *Oval mixed-cut sinhalite gem.*

Sinhalite cuts beautifully, like peridot, and is mostly found in stepped and mixed cuts. While it is suitable for jewelry, although somewhat fragile to be used in rings, it is extremely rare. Only specialist dealers will stock gemstones either for jewelry or for collectors. Its best attribute is its strong pleochroism, varying from greenish-brown and light brown to dark brown.

Buyers should beware that brown ZIRCON is sometimes passed off as the rarer and more expensive sinhalite. The best way to tell them apart is that zircon has a higher birefringence (double refraction) than sinhalite.

SINHALITE	
CHEMICAL COMPOSITION	$MgAlBO_4$
COLOR	BROWN
REFRACTIVE INDEX	1.669–1.707
RELATIVE DENSITY	3.47–3.50
HARDNESS	6.5
CRYSTAL GROUP	ORTHORHOMBIC
CLEAVAGE	DISTINCT IN TWO DIRECTIONS
FRACTURE	CONCHOIDAL
TENACITY	TOUGH
LUSTER	VITREOUS
TRANSPARENCY	TRANSPARENT TO TRANSLUCENT
DISPERSION	STRONG, .038
BIREFRINGENCE	.038
PLEOCHROISM	STRONG
LOCATION Burma (Myanmar), Russia, Sri Lanka.	

SINHALITE

Ratnapura, Sabaragamuwa province, Sri Lanka, is the type locality for sinhalite. Other sources are the Eastern Siberia region of Russia, the Warwickite Occurrence in Ontario, Canada, and the Mogok mines in Mandalay, Burma. There are unconfirmed reports of sinhalite in the Edison-Bodnar quarry, New Jersey, United States.

▲ *Translucent sinhalite crystal.*

Brilliant

Mixed

▶ *Round brilliant-cut celestine gem.*

Celestine (or celestite) is used mainly as a source of strontium for fireworks and flares. The powdered form gives off a bright crimson color if burned. It is also used as an additive to battery lead, to manufacture rubber, paint, glass, and ceramics, and to refine sugar beet. Although usually colorless, its name comes from the Latin word *caelestis*, meaning "celestial," because of the beautiful pale blue color of some crystals.

Celestine's perfect cleavage makes it very fragile, but it can be cut for the collector. Cut stones are rare and without vivid fire. A favorite with collectors is blue celestine with bright yellow sulfur. Most crystals that can be faceted come from Majunga, Madagascar.

Celestine occurs in sedimentary rocks, particularly dolostone, as cavity linings associated with BARYTE, GYPSUM, halite, anhydrite, CALCITE, DOLOMITE, and FLUORITE. It has the same structure as baryte, and forms similar crystals. The two may appear identical by ordinary methods, but a flame test can distinguish them. By scraping dust from the crystals into a gas flame, the flame's color confirms the identity of the crystal. If the flame is a pale green, it is baryte; if the flame is red, it is celestine. Celestine occurs alongside anhydrite in evaporite deposits and is often associated with sulfur in both the sedimentary environment and in volcanic areas. It occurs as a gangue mineral in hydrothermal veins with galena and SPHALERITE, also forming concretionary masses in clay and marl.

CELESTINE

Notable occurrences, besides Madagascar, include bluish crystals from Gloucestershire, England, and from Agrigento, Sicily, Italy. Large crystals are found along Lake Erie, especially at Put-in-Bay, Ohio, United States. Other US sites include Jefferson County, New York; Death Valley, California; Monroe County, Michigan; Bell's Mill, Pennsylvania; and Clay Center, Ohio. Fine celestine is also mined in Hastings County, Ontario, and Lower Saxony, Germany.

◀ *Prismatic crystals of celestine on matrix.*

CELESTINE	
CHEMICAL COMPOSITION	$SrSO_4$
COLOR	COLORLESS TO FAINT
	BLUISH WHITE OR REDDISH
REFRACTIVE INDEX	1.62–1.63
RELATIVE DENSITY	3.9–4.0
HARDNESS	3–3.5
CRYSTAL GROUP	ORTHORHOMBIC
CLEAVAGE	PERFECT
FRACTURE	UNEVEN
TENACITY	BRITTLE
LUSTER	VITREOUS
TRANSPARENCY	TRANSPARENT TO TRANSLUCENT
DISPERSION	——
BIREFRINGENCE	.01
PLEOCHROISM	NONE
LOCATION Canada, Egypt, England, France, Germany, Italy, Madagascar, Mexico, Tunisia, United States.	

Baryte (or barite) is the most common mineral source of barium. Its name comes from the Greek word *barys*, meaning "heavy," as it is unusually heavy for a non-metallic mineral. Stones are faceted only for collectors, making fine specimens. The octagonal mixed cut is especially popular. The crystals are often large, being transparent or whitish. They are normally tabular and sometimes prismatic, giving a diamond-shaped outline. They can also be fibrous or lamellar, and appear in cockscomb masses.

The mineral has many industrial uses. For example, it is powdered to produce "heavy drilling muds" used in oil wells to help prevent blowouts. Hospitals use baryte concrete and bricks to shield radioactive sources, and give "barium meals" to patients before X-rays. Baryte is also used to refine sugar, as a base for white paint, and a pigment and filler for paper.

It occurs as a vein filling and as a gangue mineral, often in lead and SILVER mines. It also accompanies ores of copper, zinc, iron, and nickel, together with CALCITE, QUARTZ, FLUORITE, DOLOMITE, and siderite. It is often confused with CELESTINE. Baryte also occurs as a replacement deposit of limestone, and as the cement in certain sandstones, sometimes with characteristic rosette-like forms called "desert roses." Hot springs also deposit the mineral.

Step

Mixed

Polished

BARYTE

The finest large crystals, up to 3 feet (1 meter) long, have been discovered in Cumbria, Cornwall, and Derbyshire, England. In the United States, reddish brown "desert roses" exist in Oklahoma and Kansas, blue crystals are found in Colorado, and other specimens have been discovered in Connecticut, New York, Pennsylvania, South Dakota, North Dakota, and Michigan. Other major locations include Transylvania, northwest Romania; Freiberg, eastern Germany; and Pribran, Czech Republic.

▲ *Yellow octagonal step-cut baryte.*

▶ *Baryte crystals.*

BARYTE	
CHEMICAL COMPOSITION	BaSO$_4$
COLOR	COLORLESS TO WHITE, TINGED YELLOW, BROWN, BLUE, GREEN, RED
REFRACTIVE INDEX	1.63–1.65
RELATIVE DENSITY	4.3–4.6
HARDNESS	2.5–3.5
CRYSTAL GROUP	ORTHORHOMBIC
CLEAVAGE	PERFECT
FRACTURE	UNEVEN
TENACITY	BRITTLE
LUSTER	VITREOUS
TRANSPARENCY	TRANSPARENT TO TRANSLUCENT
DISPERSION	WEAK
BIREFRINGENCE	.012
PLEOCHROISM	NONE
LOCATION Czech Republic, England, France, Germany, Italy, Romania, United States.	

Step

A rare lead mineral, anglesite has a sparkle that makes it popular as a gemstone, especially in its yellow variety. It was named in 1832 after the Isle of Anglesey, Wales, where it was discovered at Porys Mine, and anglesite probably remains the best-known Welsh mineral among collectors. It is also valued as a source of lead, used in consumer items such as car batteries and computer screens.

Anglesite occurs in weathered deposits of lead ore, formed by the oxidation of primary deposits of galena. It is found as coatings or crusts on galena, often as masses surrounding a galena core, and as crystal aggregates filling the cavities of galena. CERUSSITE is often confused with anglesite, and other associated minerals include wulfenite, SMITHSONITE, mimetite, pyromorphite, and limonite. Anglesite is part of the baryte group, having the same structure as baryte with similar crystals that are normally flat blades or pendent columns jutting from the rock.

▲ *Anglesite as crystal aggregate.*

▶ *Anglesite crystals.*

ANGLESITE

Some of the best gems come from Tsumeb, Namibia, where high-luster, diamond-shaped deposits measuring up to 20 inches (50 centimeters) have been recovered. Other sites include Broken Hill, Australia; Leadhills, Scotland; Cumbria, England; North Rhine-Westphalia, Germany; Cagliari province, Sardinia; Oujda, Morocco; and various US locations, such as Chester County, Pennsylvania; Dividend, Utah; and Bingham, New Mexico.

ANGLESITE	
CHEMICAL COMPOSITION	$PbSO_4$
COLOR	COLORLESS TO WHITE, SOME TINGED YELLOW, GRAY, GREEN, BLUE
REFRACTIVE INDEX	1.88–1.89
RELATIVE DENSITY	6.32–6.4
HARDNESS	2–3
CRYSTAL GROUP	ORTHORHOMBIC
CLEAVAGE	BASAL, GOOD; PRISMATIC, DISTINCT
FRACTURE	CONCHOIDAL
TENACITY	BRITTLE
LUSTER	ADAMANTINE
TRANSPARENCY	TRANSPARENT TO TRANSLUCENT
DISPERSION	.044
BIREFRINGENCE	.017
PLEOCHROISM	NONE

LOCATION Australia, Canada, Germany, Italy, Mexico, Morocco, Namibia, United Kingdom, United States.

▶ *Round brilliant-cut anglesite gem.*

Gypsum is a hydrated calcium sulfate, which is the most common sulfate mineral. The name is derived from the Greek word *gypsos*, meaning "plaster." Gypsum can be scratched with a fingernail and easily cut by a knife. It has several distinctive varieties that are used as gemstones, especially the fine-grained alabaster. **Satin spar**, fibrous gypsum and white gypsum have a silky luster and are often cut as cabochons and polished to produce the cat's-eye effect. Soft **selenite**, which is colorless and transparent, is sometimes cut. Popular with mineral collectors are the beautiful "desert roses," the "swallow-tail" twins, and stellate, or star-like, forms.

Gypsum is also used in plasters, such as plaster of Paris, fertilizers, Portland cement, and as a filler in paper, paints, and crayons. It is the most common evaporite, a deposit left after the evaporation of water. Gypsum is discovered in massive beds in sedimentary rocks with limestones and shale. It is usually produced by the secondary hydration of anhydrite. Gypsum is also associated with CALCITE, sulfur, QUARTZ, DOLOMITE, halite, and clay. Some gypsum is deposited by the evaporation of saltwater or crystallizes in dry lakes as soft, translucent crystals. It also occurs as free crystals in clay, as cap rock on salt domes, and in volcanic areas. **Alabaster**, both massive and fine-grained, is used for statues and ornamental carvings. This same softness, however, causes it to break easily and weather badly. It is generally translucent with a white, pinkish, or brownish color.

Cabochon

GYPSUM

Most alabaster comes from Volterra in Tuscany, Italy, and from Derbyshire, Staffordshire, and Nottinghamshire in England. Pink alabaster mines exist in the Vale of Glamorgan, Wales. Other countries with alabaster sites include Spain, Iran, and Pakistan. The "alabaster" used in ancient Egypt and Rome for vases, tombs, and other objects was, in fact, marble (calcium carbonate).

Varieties of gypsum occur in several US states, such as Arizona, California, Utah, Colorado, Oklahoma, New Mexico, Ohio, Michigan, Virginia, Kentucky, Pennsylvania, Washington, and New York. Other important mines are in Nova Scotia, Ontario, and New Brunswick, in Canada, and Languedoc-Rousillon and Alsace, France.

GYPSUM

CHEMICAL COMPOSITION	$CaSO_4.2H_2O$
COLOR	COLORLESS TO WHITE, SOMETIMES SHADES OF YELLOW, GRAY, RED, AND BROWN
REFRACTIVE INDEX	1.52–1.53
RELATIVE DENSITY	2.3
HARDNESS	2
CRYSTAL GROUP	MONOCLINIC
CLEAVAGE	ONE PERFECT, TWO GOOD
FRACTURE	CONCHOIDAL
TENACITY	SECTILE
LUSTER	VITREOUS
TRANSPARENCY	TRANSPARENT TO TRANSLUCENT
DISPERSION	STRONG
BIREFRINGENCE	.010
PLEOCHROISM	NONE

LOCATION Australia, Canada, France, Germany, Italy, Kazakhstan, Mexico, Russia, United Kingdom, United States.

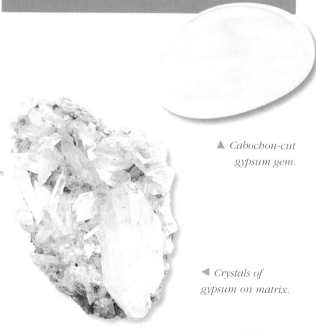

▲ *Cabochon-cut gypsum gem.*

◀ *Crystals of gypsum on matrix.*

Brilliant

Step

Mixed

Scheelite appeals to collectors of rare and unusual gems. In its raw state, it forms perfect tetragonal dipyramidal crystals that look like octohedrons. It is also attractive when faceted and polished. Scheelite generally accompanies wolframite in pegmatites and high-temperature hydrothermal veins. Associated minerals are CASSITERITE, molybdenite, FLUORITE, and TOPAZ. It also occurs in contact metamorphic deposits together with VESUVIANITE, axinite, GARNET, and wollastonite.

Named for Carl Wilhelm Scheele (1742–86), the Swedish chemist who discovered tungsten, scheelite is one of the major ores of tungsten and the biggest source of tungsten in the United States. Because it is quite soft, scheelite is rarely used for jewelry. However, faceted gems of scheelite are quite attractive, and specimens from the Pingwu mine in Sichuan province, west-central China, have an attractive yellow to orange color and good fire that appeal to collectors. When cutting, it is important to be careful, and a perfect polish may not be possible.

Scheelite fluoresces a vivid blue under shortwave ultraviolet light. In fact, some prospectors find the rock by searching at night using fluorescent lamps.

► *Pendeloque-cut scheelite from Sri Lanka.*

SCHEELITE

CHEMICAL COMPOSITION	$CaWO_4$
COLOR	WHITE, SOMETIMES SHADES OF YELLOW, GREEN, BROWN OR RED
REFRACTIVE INDEX	1.92–1.93
RELATIVE DENSITY	5.6–6.1
HARDNESS	4.5–5.0
CRYSTAL GROUP	TETRAGONAL
CLEAVAGE	PYRAMIDAL, DISTINCT
FRACTURE	CONCHOIDAL
TENACITY	———
LUSTER	VITREOUS
TRANSPARENCY	TRANSPARENT TO TRANSLUCENT
DISPERSION	MODERATELY HIGH, .026
BIREFRINGENCE	MEDIUM, .014–.016
PLEOCHROISM	DICHROIC YELLOW: POSITIVE, BROWN TO ORANGE; COLORLESS: NONE

LOCATION Australia, Brazil, China, Finland, France, Italy, Mexico, Sri Lanka, Sweden, United Kingdom, United States.

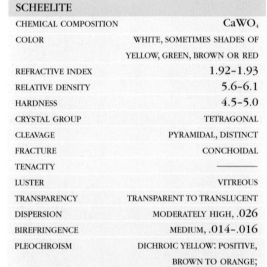

◄ *Scheelite crystals on matrix.*

► *Faceted scheelite.*

SCHEELITE

Other mines exist in Yukon Territory, Canada; Cornwall, England; Inyo County and San Bernardino County, California; and Okanogan County, Washington, United States.

Cushion

Pendeloque

Baguette

A phosphate mix of sodium and aluminum, brazilianite was discovered in 1944 at Minas Gerais, southeast Brazil, and at first incorrectly assumed to be chrysoberyl. It is a member of the phosphate group and has often been confused with TOPAZ, APATITE, and BERYL. The US mineralogist F. H. Pough soon identified the mineral and, with co-author E. P. Henderson, named it after its country of discovery. Jewelers and collectors prize the gemstone, which is a rare, unusual, and beautiful crystal with a striking yellowish-green color. It is considered to be more precious than the phosphate mineral apatite, but brazilianite is too new to mineral markets to be very popular worldwide. The crystals, brittle and fragile, chip easily, breaking along the direction of cleavage. Although lapidaries must take great care in cutting, brazilianite has been faceted as cushion, baguette, pendeloque, brilliant, and oblong step-cut gems. Most gems are under 5 carats, but some large stones have been cut, including the first specimens of 19 and 23 carats.

▲ *Round cushion-cut brazilianite gem.*

▶ *Brazilianite crystals on matrix.*

BRAZILIANITE

CHEMICAL COMPOSITION	NaAl₃[(OH)₂/PO₄]₂
COLOR	YELLOW GREEN
REFRACTIVE INDEX	1.60–1.62
RELATIVE DENSITY	2.99
HARDNESS	5.5
CRYSTAL GROUP	MONOCLINIC
CLEAVAGE	GOOD
FRACTURE	CONCHOIDAL
TENACITY	BRITTLE
LUSTER	VITREOUS
TRANSPARENCY	TRANSPARENT TO TRANSLUCENT
DISPERSION	WEAK, .014
BIREFRINGENCE	.02
PLEOCHROISM	WEAK

LOCATION Brazil, United States.

BRAZILIANITE

Brazilianite is one of the hardest phosphate minerals. Its associated minerals include mica, QUARTZ, FELDSPAR, muscovite, BERYL, apatite, and TOURMALINE. It occurs in phosphate-rich pegmatites as druse crystals that are longitudinally striated, a feature that helps in the selection of cuts. The crystals may have a multifaced termination, and be either long and pointed or short and columnar. Massive examples are extremely rare.

The major sources of the stone are near Conselheiro Peno and Linópolis, Minas Gerais, Brazil. In 1947, the gem was also discovered in the US state of New Hampshire at the G. H. Smith Mine at Newport and the Palermo Mine at Groton.

▶ *Green marquise-cut brazilianite gem.*

AMBLYGONITE

Brilliant

Mixed

An aluminum and lithium fluophosphate, amblygonite was first discovered in 1817 in Saxony, east-central Germany. Its name, from the Greek words *amblys* ("blunt") and *goni* ("angle"), refers to the mineral's four angles of cleavage. Used as a common source of lithium and phosphorus, amblygonite is also popular in porcelain enamels and increases opacity in glass dinnerware. The best examples of its yellow variety are turned into gemstones, with some samples clean enough to be faceted. The presence of lithium can be determined by a gas-flame test of powdered amblygonite, yielding a bright red flame. Strong sulfuric acid is used to recover the lithium as sulfate.

Amblygonite is rare and its good crystal forms, shaped as narrow prisms, are even rarer. The mineral occurs in granite pegmatites in compact, cleavable masses, usually white. These veins are alongside other phosphates and lithium minerals, such as SPODUMENE, LEPIDOLITE, TOURMALINE, and ALBITE. Albite and quartz are often mistaken for amblygonite. Some of amblygonite's lithium is replaced by sodium and some of its fluorine by hydroxyl. Although amblygonite and **montebrasite** have the same structure and are normally considered the same, amblygonite is richer in fluorine and montebrasite in hydroxide. Other members of the amblygonite group are natramblygonite, natromontebrasite, and tavorite.

AMBLYGONITE

Amblygonite's type locality is Penig, Chemnitz, Saxony, east-central Germany. Amblygonite mines are found in Minas Gerais (particularly around Araçuaí), southeast Brazil; Sakangyi, Mogok, Mandalay, central Burma (Myanmar); Montebras, Limousin, central France; Bernic Lake, Manitoba, central Canada. Major US locations include Maricopa County, Arizona; Taos County, New Mexico; Riverside County, California; and the Black Hills of South Dakota. Gem-quality amblygonite is mined mostly in Brazil and Burma (Myanmar).

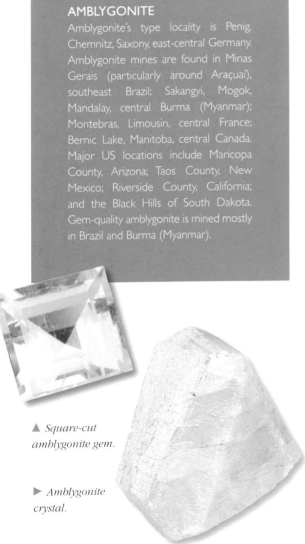

▲ *Square-cut amblygonite gem.*

▶ *Amblygonite crystal.*

AMBLYGONITE	
CHEMICAL COMPOSITION	(Li,Na)AlPO$_4$(F,OH)
COLOR	WHITE, PALE GREEN, PALE BLUE, PINK, PALE YELLOW
REFRACTIVE INDEX	1.58–1.61
RELATIVE DENSITY	3.0–3.1
HARDNESS	5.5–6
CRYSTAL GROUP	TRICLINIC
CLEAVAGE	ONE PERFECT, TWO GOOD, ONE DISTINCT
FRACTURE	UNEVEN
TENACITY	——
LUSTER	VITREOUS TO GREASY OR PEARLY
TRANSPARENCY	SUBTRANSPARENT TO TRANSLUCENT
DISPERSION	WEAK
BIREFRINGENCE	.020–.027
PLEOCHROISM	NONE

LOCATION Brazil, Burma (Myanmar), Canada, France, Germany, Namibia, Sweden, United States.

Apatite is a calcium phosphate that is fashioned into colorful gemstones, although the mineral is too soft to gain a worldwide importance as a gem (it can be scratched with a steel knife). While jewelry is fashioned, normally cut into cabochons, apatite is not used in rings because of its poor durability. The mineral's phosphorous is used to manufacture fertilizers. Apatite also contains calcium, iron, chlorine, and fluorine.

Apatite occurs as fluorapatite (the principal constituent in teeth and bones), chloroapatite, and hydroxylapatite. The members of the apatite group, which also include mimetite and vanadinite, are the most common phosphorus-bearing minerals.

Apatite's name was created in Germany from the Greek word *apatao*, meaning "to deceive," because its many colors create confusion with other gemstones, such as BERYL and OLIVINE. The array of apatite colors include white, yellow, green, violet, blue, brown, and gray. A Spanish yellow-green variety is called "asparagus stone" because it resembles the vegetable's color.

Present in many countries, apatite is normally found as small crystals that are usually prismatic or tabular. It is an accessory mineral in a wide range of igneous rocks. Apatite is a main constituent of fossil bones and other organic matter in sedimentary rocks. Large crystals occur in pegmatites and some high-temperature hydrothermal veins. Apatite also occurs as larger crystals in metamorphic rocks, especially in metamorphosed limestones and skarns. This variety is known as **collophane** or collophanite.

Step

Baguette

Cushion

Cabochon

APATITE	
CHEMICAL COMPOSITION	$Ca_5(PO_4)_3(F,Cl,OH)$
COLOR	GREEN, GRAY-GREEN, WHITE, BROWN, YELLOW, BLUISH, REDDISH
REFRACTIVE INDEX	1.63–1.64
RELATIVE DENSITY	3.1–3.3
HARDNESS	5
CRYSTAL GROUP	HEXAGONAL
CLEAVAGE	INDISTINCT
FRACTURE	CONCHOIDAL, UNEVEN
TENACITY	BRITTLE
LUSTER	VITREOUS TO SUBRESINOUS
TRANSPARENCY	TRANSPARENT TO TRANSLUCENT
DISPERSION	.013
BIREFRINGENCE	.003
PLEOCHROISM	WEAK TO STRONG

LOCATION Brazil, Burma (Myanmar), Canada, Czech Republic, Germany, India, Madagascar, Mexico, Norway, Russia, Spain, Sri Lanka, United Kingdom, United States.

APATITE

Major deposits of apatite are found at Durango, Mexico (well-known for specimens of yellow variety); Bancroft, Ontario, Canada; Cornwall, England; Mogok, central Burma (Myanmar); Vestfold, Norway; New York, USA; Kola Peninsula, Russia; Bavaria, Germany; and Lappland, Sweden.

▲ *Green prismatic crystal of apatite.*

▶ *Oval cushion-cut apatite gem.*

▲ *Art Nouveau turquoise, gold and enamel necklace, made in c. 1900. The centerpiece of the necklace is a cabochon-cut turquoise.*

▼ *Cabochon-cut turquoise gem.*

A hydrated phosphate of copper and aluminum, turquoise is prized as a gemstone whose intense blue color is often mottled with veinlets of brown limonite or black manganese oxide ("spiderweb" turquoise). The name comes from the French *turquoise*, meaning "Turkish," since it arrived in Europe through Turkey, originating in Iran. It has been mined in Iran for more than 3,000 years, although the earliest source of turquoise may have been Egypt's Sinai Peninsula.

Turquoise is cut as cabochons, or as flat pieces for inlaid work. It is also used as beads and cameos. Irregular pieces are set in mosaics. The most renowned turquoise jewelry in the United States has long been produced by Native Americans, especially the Zuni and Navajo peoples. They create turquoise and silver jewelry pieces like bracelets, necklaces, rings, earrings, pendants, brooches, and belt buckles. Other inlays are often included, such as SHELL, CORAL, LAPIS LAZULI, and MALACHITE. The overall piece is then sanded and polished.

Jewelers must take special care, because the friction of polishing can turn turquoise an unattractive green color. The stone is also sensitive to sunlight, soap, water, dry air, perspiration, and grease. Some porous specimens, especially in the United States, may require an impregnation with resin or wax in order to resist fading and cracking. "Turquoise matrix," a natural aggregate of turquoise with limonite or other minerals, is also used for gemstones. Exceptional white or brown matrix stones exist in the southwest United States. In 1972, an imitation "Gilson turquoise" was created in France, made up of blue

TURQUOISE

Turquoise is a secondary mineral occurring in veins in association with aluminous, igneous or sedimentary rocks that have undergone considerable alteration, usually in arid regions. It is distinguished from CHRYSOCOLLA by its greater hardness. It can be confused with LAZULITE, variscite, and wardite. A very small amount of turquoise replaces bones and other fossil materials. The crystals are rare and minute, usually massive, granular to cryptocrystalline habits as reniform or encrusting masses, or in veins. Distinct crystals have only been found in Campbell County, Virginia, United States.

Other US turquoise occurs in the Mojave Desert of California, the Cerrillos Hills near Sante Fe in New Mexico, and the states of Arizona, Nevada, Utah, and Colorado. Many of these deposits were mined centuries ago by Native Americans. Sky-blue turquoise occurs in Iran and a green variety in Tibet. Additional mines are in Cornwall in England, Victoria and Queensland in Australia, Siberia in Russia, and France, Germany, Chile, Egypt, Tibet, and China.

angular pieces on a white background. Other imitations have included stained limestone, howlite, CHALCEDONY, glass, enamel, and fossil bone and tooth.

Turquoise was one of the first gemstones ever mined, perhaps dating back to 6000 BC in Egypt's Sinai Peninsula. It has been used for thousands of years as jewelry by the ancient Egyptians, who buried fine pieces with mummies, and by Native Americans, who combined turquoise with silver. The Aztecs fashioned elaborate turquoise masks. In North America, the Zuni people of New Mexico have created striking turquoise jewelry set in silver, once believing these protected them from demons. The Navajo believed the blue stones were pieces that had fallen from the sky, and Apache warriors wore it in the belief that it improved their hunting prowess. It was also worn by other peoples in the belief that the stone's color change could indicate illness or danger.

Bead

Cabochon

Cameo

◄ *Turquoise and diamond necklace made in c. 1880. Designed as a snake, the head supports a heart set with cabochon-cut turquoise. The eyes are cabochon-cut garnets, framed by rose-cut diamonds.*

TURQUOISE	
CHEMICAL COMPOSITION	$CuAl_6(PO_4)_4$ $(OH)_8.4H_2O$
COLOR	SKY-BLUE, BLUE-GREEN TO GREENISH GRAY
REFRACTIVE INDEX	1.61–1.65
RELATIVE DENSITY	2.6–2.8
HARDNESS	5–6
CRYSTAL GROUP	TRICLINIC
CLEAVAGE	NONE
FRACTURE	CONCHOIDAL
TENACITY	BRITTLE
LUSTER	WAXY IF MASSIVE; VITREOUS AS CRYSTALS
TRANSPARENCY	NEARLY OPAQUE
DISPERSION	NONE
BIREFRINGENCE	.04
PLEOCHROISM	WEAK
LOCATION Australia, Chile, China, Egypt, England, France, Germany, Iran, Mexico, Russia, Tibet, Turkmenistan, United States.	

▲ *Two masses of turquoise showing botryoidal habit.*

LAZULITE

Cabochon

A rare aluminum phosphate, lazulite has a luster akin to glass and is used as a gemstone. Its name derives from the Arabic word *azul* meaning "heaven," because of its often sky-blue color, and the Greek word *lithos* for "stone." The colors range from a whitish blue to deep blue.

Lazulite is a rare ornamental stone with mostly dull crystals, which have a bipyramidal form that is normally flattened into a tabular shape. Gem-quality pieces that can be cut are rare. Tiny crystal fragments can be found that are semitranslucent or opaque and well formed. If more than a few carats, they can fetch exceptional pieces. They are often carved and polished into decorative stones and beads that are much sought after because of their rich color. A cabochon's polished dome can reveal a beautiful blue and white mottling effect.

Lazulite occurs in pegmatites and quartz veins and in quartzites. Associated minerals include KYANITE, CORUNDUM, RUTILE, SILLIMANITE, GARNET, and BRAZILIANITE. Scorzalite, an iron-rich mineral, is similar to lazulite, being less transparent, darker, and more dense. When massive, lazulite is difficult to distinguish from other blue minerals and may be confused with TURQUOISE, variscite, and the similarly named LAZURITE (lapis lazuli), and AZURITE.

LAZULITE

Lazulite is the official gemstone of Yukon Territory, Canada. The gem's type localities are Werfen, Salzburg, central Austria, and Freßnitzgraben, Styria, southeast Austria. The Graves Mountain Mine in Lincoln County, Georgia, United States, produces a striking mixture of deep blue lazulite, light blue kyanite, and gray quartzite. Other sources of lazulite include Skåne, Sweden; Bavarian Forest, Germany; Tuscany, Italy; and the US states of California, Connecticut, New Hampshire, and North Carolina.

◀ *Lazulite crystal in matrix.*

▼ *Polished oval lazulite from Pakistan, 7.6 carats.*

LAZULITE	
CHEMICAL COMPOSITION	$(Mg,Fe)Al_2(PO_4)_2(OH)_2$
COLOR	DEEP AZURE-BLUE
REFRACTIVE INDEX	1.61–1.64
RELATIVE DENSITY	3.0–3.1
HARDNESS	5–6
CRYSTAL GROUP	MONOCLINIC
CLEAVAGE	PRISMATIC, INDISTINCT
FRACTURE	UNEVEN
TENACITY	BRITTLE
LUSTER	VITREOUS
TRANSPARENCY	TRANSLUCENT
DISPERSION	NONE
BIREFRINGENCE	———
PLEOCHROISM	STRONG
LOCATION Austria, Brazil, Germany, Italy, Pakistan, Russia, Slovak Republic, Sweden, United States.	

Peridot is the gem-quality variety of the mineral OLIVINE. A rock-forming mineral, olivine occurs in silica-poor igneous rocks, such as basalts, gabbro, troctolite, and peridotite. The crystals mostly project as granular olivines, but also in tabular forms. Peridot's shades of green are caused by the presence of iron, and it often has a greasy luster. The mineral is named for the French word *peritot*, meaning "gold," because the mineral can vary toward this color.

Peridot crystals are fashioned in the trap-cut style, and also faceted as cabochon, pende-loque, step, table, brilliant, rose, and mixed cuts. Peridot is too soft to be used as gemstones in rings or bracelets, but is popular for earrings, pendants, clasps, and brooches. Small crystals are relatively common, but large, good-quality ones are rare. In some locations, finding a stone of 5 carats causes excitement. The Smithsonian Institution in Washington, D.C., has a cut stone of 310 carats.

◀ *Salamander brooch by Van Cleef and Arpels, set with peridots, red garnet eyes, and diamond details.*

Step

Baguette

PERIDOT

The main source of peridot in the ancient world was Topazos Island (now Zebirget or St John's Island) in the Egyptian Red Sea, and the stones were called topazios. They were used for carved talismans in ancient Egypt. In the Middle Ages, Europeans brought the crystals back from the Crusades to decorate church plates and robes. In the 19th century, the mines on Zebirget Island produced millions of dollars' worth of peridot; today they lie almost silent. US sources include the San Carlos Apache Reservation, Arizona; Kilbourne Hole, New Mexico; and Salt Lake Crater, Hawaii (beach pebbles). Other sources include Pyaung gaung, Mogok, Burma (Myanmar), and Northwest Frontier Province, Pakistan.

▲ *Crystal fragment of peridot.*

Cushion

PERIDOT	
CHEMICAL COMPOSITION	$(Mg, Fe)_2SiO_4$
COLOR	OLIVE GREEN TO YELLOWISH GREEN
REFRACTIVE INDEX	1.64–1.69
RELATIVE DENSITY	3.34
HARDNESS	6.5
CRYSTAL GROUP	ORTHORHOMBIC
CLEAVAGE	INDISTINCT
FRACTURE	CONCHOIDAL
TENACITY	BRITTLE
LUSTER	VITREOUS TO GREASY
TRANSPARENCY	TRANSPARENT TO TRANSLUCENT
DISPERSION	MEDIUM, .02
BIREFRINGENCE	.036
PLEOCHROISM	DISTINCT
LOCATION Burma (Myanmar), Canada, Egypt, Norway, Pakistan, United States.	

▲ *Three faceted peridot gems.*

PHENAKITE

Brilliant

Mixed

▶ *Triangular fancy-cut phenakite gem.*

Phenakite (or phenacite) is a rare beryllium mineral that is not often used as a gemstone. Its name derives from the Greek word *phenakos*, meaning "deceiver," a name acquired because of its close similarity to quartz, especially rock crystal. Indeed, phenakite was initially thought to be a variety of quartz, but its crystal twinning is distinct. It has also been confused with diamond, sapphire, topaz, and beryl.

▲ *Cabochon-cut phenakite gem.*

Phenakite crystals are often rhombohedral and sometimes prismatic, showing wedge-shaped ends. Clear, hard specimens can be turned into valued gemstones despite their lack of color and fire. If cut properly, they display a silvery look. Transparent crystals, which are bright and hard, are the only ones that are faceted, normally as brilliant cuts but also as mixed cuts. Translucent white phenakite can be cut to display a cat's-eye effect.

Phenakite and dioptase are silicate minerals with trigonal symmetry. Phenakite occurs in cavities in granites and in granite pegmatites in association with beryl, chrysoberyl, topaz, apatite, quartz, albite, and mica. It also occurs in metamorphic rocks carrying beryl and in hydrothermal veins. The pink, yellow, and brown specimens may fade under sunlight within months of being extracted.

PHENAKITE

A rich site for phenakite is the Sverdlovskaya region in the eastern foothills of the Ural Mountains, southern Russia, where the mineral was first discovered in emerald mines in the early 19th century. Good-sized colorless crystals have also been found in Minas Gerais, Brazil, and long columnar ones at Kragerø, Norway. A colorless fragment of 1,470 carats was recently discovered in Sri Lanka and cut into a 569-carat oval and other smaller stones. US sources include El Paso County, Colorado, and Carroll County, New Hampshire.

◀ *Phenakite crystals on matrix.*

PHENAKITE	
CHEMICAL COMPOSITION	Be2iO4
COLOR	COLORLESS, WHITE, YELLOW, PINKISH, BROWN
REFRACTIVE INDEX	1.65–1.67
RELATIVE DENSITY	3
HARDNESS	7.5–8
CRYSTAL GROUP	TRIGONAL
CLEAVAGE	POOR
FRACTURE	CONCHOIDAL
TENACITY	BRITTLE
LUSTER	VITREOUS
TRANSPARENCY	TRANSPARENT TO TRANSLUCENT
DISPERSION	WEAK, .015
BIREFRINGENCE	.01
PLEOCHROISM	DISTINCT IN COLORED STONES
LOCATION Brazil, Norway, Russia, Sri Lanka, United States.	

Kornerupine is a rare mineral and gemstone. The intense color resembles that of the emerald, and it has the inclusions of that gem. Kornerupine is distinguished by its pleochroic colors that change from a yellowish green, through blue, to brownish red as the stone is turned. It is mostly a collector's stone. Few retail jewelers stock the stones, but they are becoming more popular for jewelry, being moderately priced and providing a gemstone for everyday wear. Kornerupine is cut on the green axis for the highest value. Cat's-eye stones are polished as cabochons. It is named after Andreas N. Kornerup (1857–81), a Danish geologist who explored Greenland, where the mineral was discovered at Fiskenaesset in 1884.

A metamorphic mineral, kornerupine is a complex magnesium aluminum boro-silicate, whose crystals have long prismatic forms or occur as rounded grains often found in gravel deposits behind rocks and at bends of rivers. It is deposited with other gems like sapphire, chrysoberyl, ruby, topaz, garnet, zircon, diopside, andalusite, spinel, sphene, and iolite.

Cushion

Step

Baguette

◀ *The crystal is usually cut on the green axis but can also be blue.*

KORNERUPINE	
CHEMICAL COMPOSITION	Mg₄Al₆[(O, OH)₂/BO₄/(SiO₄)₄]
COLOR	GREEN, GREENISH BROWN, BLUE, BROWNISH RED
REFRACTIVE INDEX	1.66–1.68
RELATIVE DENSITY	3.28–3.35
HARDNESS	6.5–7
CRYSTAL GROUP	ORTHORHOMBIC
CLEAVAGE	GOOD
FRACTURE	CONCHOIDAL
TENACITY	BRITTLE
LUSTER	VITREOUS
TRANSPARENCY	TRANSPARENT
DISPERSION	MODERATE, .018
BIREFRINGENCE	.01
PLEOCHROISM	STRONG

LOCATION Australia, Burma, Canada, Greenland, Kenya, Madagascar, Sri Lanka, Tanzania.

KORNERUPINE

Fiskenaesset, Nuuk, Greenland, is the type locality of kornerupine, but specimens are rare and are not of gem quality. The best gemstones, including rare cat's-eyes, are found in Mogok, Mandalay, central Burma (Myanmar). Some of the finest colors of kornerupine gems are mined at Betroka, Madagascar. Other sources include Ratnapura in southwest Sri Lanka, and the Harts Range in Northern Territory, Australia.

▶ *Round mixed-cut kornerupine gem.*

127

DIOPTASE

Brilliant

Cabochon

Despite its beautiful emerald-like green color, dioptase is not as highly valued as EMERALD because of its relatively low hardness of 5, which makes it vulnerable to damage when set in jewelry. It is sometimes known as the "emerald of the poor" for this reason. It was named in 1797 by mineralogist R. J. Haüy (1743–1822) from the Greek *dia* for "through" and *optomai* "to see" – a reference to the visibility of its internal cleavage planes. Other names for dioptase include "copper emerald" and "achrite."

Dioptase is highly valued by mineral collectors. Its crystals usually take the form of short, six-sided prisms, often terminated by rhombohedra. It also occurs in massive form. It is not a common mineral, but is found in the oxidized, weathered parts of copper sulfide deposits and with copper minerals such as CHRYSOCOLLA. It is also found in association with DOLOMITE, CALCITE, CERUSSITE, and limonite. It grows in cracks in rocks and in the cavities of druses. In the field, mineralogists recognize dioptase by the distinguishing features of its color, crystal form, and association with copper minerals. Its streak is green.

Only the clear ends of large crystals are used for faceting. Cut stones often have a pearly appearance caused by reflection from tiny internal cleavage cracks. The perfect cleavage makes it difficult to fashion step- and table-cuts. Fine-grained masses of dioptase may be made into cabochons.

▶ *Dioptase crystals on matrix.*

DIOPTASE

CHEMICAL COMPOSITION	$CuSiO_2(OH)_2$
COLOR	EMERALD-GREEN
REFRACTIVE INDEX	1.65–1.71
RELATIVE DENSITY	3.28–3.35
HARDNESS	5
CRYSTAL GROUP	TRIGONAL
CLEAVAGE	PERFECT
FRACTURE	CONCHOIDAL TO UNEVEN
TENACITY	BRITTLE
LUSTER	VITREOUS
TRANSPARENCY	TRANSPARENT TO TRANSLUCENT
DISPERSION	STRONG, .028
BIREFRINGENCE	.053
PLEOCHROISM	WEAK

LOCATION Chile, Democratic Republic of Congo, Republic of the Congo, Kazakhstan, Mexico, Namibia, United States.

DIOPTASE

Major dioptase deposits exist on the type site of Mount Altyn-Tyube, Kazakhstan. In Africa, large crystals are found in Otavi, Namibia, and the Pool region of the Republic of the Congo. Other sources include the Shaba province of the Democratic Republic of Congo. In the Americas, dioptase is found in Chile and Argentina. Pinal County, Arizona, is the major source in the United States.

▶ *Cushion-cut dioptase from South Africa, 1.40 carat.*

An unusual and beautiful gem, sphene (or **titanite**) is rarely seen either in jewelry or in gem collections. Sphene has a fire greater than diamond, but it is rather soft and brittle so must be handled with extreme care. The name "titanite" derives from the fact that it is an ore of titanium, while the name "sphene" is a reference to its wedge-shaped crystals.

Sphene is widely distributed as an accessory mineral, particularly in coarse-grained igneous rocks, such as syenite, nepheline syenite, diorite, and granodiorite. It occurs similarly in schists or gneisses and in some metamorphosed limestones. Sphene is appealing because of its high birefringence and adamantine luster, as well as for the strong pleochroism apparent in the colored stones. Collectors appreciate its crystals, in which twinning is common, as well as the striking appearance of the faceted gems. It is best cut into brilliants or mixed cuts. They are usually small in size, because gems of more than 2 carats are extremely rare. Although sphene's softness makes it unsuitable for use in rings, it can make beautiful earrings and pendants, which catch the light wonderfully.

Cushion

Step

◀ *Bladed sphene crystals on matrix.*

SPHENE	
CHEMICAL COMPOSITION	CaTiSiO$_5$
COLOR	BROWN AND GREENISH YELLOW, SOMETIMES GRAY OR NEARLY BLACK
REFRACTIVE INDEX	1.885–2.05
RELATIVE DENSITY	3.4–3.56
HARDNESS	5.5
CRYSTAL GROUP	MONOCLINIC
CLEAVAGE	PRISMATIC, DISTINCT
FRACTURE	CONCHOIDAL
TENACITY	BRITTLE
LUSTER	RESINOUS TO ADAMANTINE
TRANSPARENCY	TRANSPARENT TO TRANSLUCENT, OCCASIONALLY NEARLY OPAQUE
DISPERSION	STRONG, .051
BIREFRINGENCE	.13
PLEOCHROISM	STRONG IN DEEP-COLORED STONES

LOCATION Canada, Germany, Italy, Norway, Pakistan, Russia, Sweden, United Kingdom, United States.

SPHENE

The major sources of sphene include Gilgit, northern Pakistan; Eifel Mountains, Rhineland-Palatinate, western Germany; Mount Vesuvius, Naples, southern Italy; Kola Peninsula, northern Russia; Västmanland, Sweden; Haliburton County and Hastings County, Ontario, Canada; and Orange County, New York, and Sussex County, New Jersey, USA.

▶ *Cushion-cut sphene gem.*

Brilliant

Cushion

Baguette

Mixed

Zircon

▼ *The zircon-cut requires additional facets on the pavilion of the stone.*

Zircon is a zirconium silicate well known for its popular gem-quality stones. The name derives from the Arabic word *zargun*, meaning "golden color." Zircon is a source of the metal zirconium, which took its name from the mineral and is now used in the production of nuclear reactors.

Gem-quality zircon crystals have an adamantine luster and are normally found as pebbles in alluvial deposits. In pegmatites, the crystals can reach a considerable size, but most are small, usually prismatic, with bipyramidal terminations. Twinning is common, giving knee-shaped twins. Impurities produce blue, red, brown, green, yellow, and orange varieties. Most zircons contain traces of radioactive uranium or thorium that substitute for zirconium and will eventually break down the crystal structure. These decayed stones, often green, are known as "low zircon."

Sri Lanka has produced gem-quality zircon for more than 2,000 years, and zircon jewelry has been fashioned there and in India for centuries. The gemstone, however, did not become fashionable in

ZIRCON

Zircon is one of the most widely distributed accessory minerals in igneous rocks such as granite, syenite, and nepheline syenite. It occurs in hydrothermal veins in association with minerals such as QUARTZ, FLUORITE, DOLOMITE, PYRITE, SPHALERITE, BARITE, and chalcopyrite. Zircon is also found in metamorphic rocks, such as schists and gneisses, and becomes concentrated as a detrital mineral in beach and river sands. It is found also in sandstones bearing GOLD.

Zircon is widely distributed. Pailin, western Cambodia, is the best source for gem-quality blue zircon. Other top sites for high-quality zircon gems include the Mogok mines in Mandalay, central Burma (Myanmar); Ratnapura, Sri Lanka; and Kanchanaburi, Thailand. Well-formed red crystals come from the Auvergne, France. Brown crystals are found at Arendal, Norway, and near-white rolled pebbles exist in Tanzania. Other sources include Victoria and the beach sands of New South Wales, Australia; Arendal and Langesund, Norway; Kola Peninsula and the gold gravels of the Urals, Russia; Haliburton County and Hastings County, Ontario, Canada; Oberpfalz, Bavaria, and the Eifel Mountains, Rhineland-Palatinate, western Germany; Honshu, Japan; Värmland, Sweden; and at Litchfield, Maine, United States. US deposits include Auburn and Greenwood, Maine; El Paso County, Colorado; Sussex County, New Jersey; and Orange County, New York. Deposits of zircon have also been found on the Moon.

Gem-quality zircon can be found in a number of colors. Impurities produce blue, red, brown, green, yellow, and orange varieties. It can also be heat-treated to create a variety of colors.

Zircon crystals on matrix.

western countries until the 1920s. It can be heat-treated to create a variety of colors. Most colorless, blue, and golden stones used in jewelry are heat-treated brown zircons from Thailand, Vietnam, Sri Lanka, and Cambodia. Blue zircon reheated in oxygen will change to a golden yellow color. Artificially colored zircons may fade in sunlight over time.

The gem's luster, hardness, and many colors make it a popular stone for rings. Reddish-brown varieties have been called **hyacinth** or **jacinth**. It is most popular as blue, brown, golden, or colorless stones. The colorless and pale stones are usually cut as round and oval brilliants. Reddish-brown hyacinth is also fashioned into step, mixed, baguette and cushion cuts. Transparent zircon is used for gemstones, and colorless zircon resembles diamond and has been sold as such, especially when fashioned as rose cuts in earlier times. Like diamond, it can break up white light into spectral colors. Zircon differs from diamond by its double refraction and, because of its extreme brittleness, by the wear and chipping on the faceted edges of cut stones. Zircon should not be confused with the artificial stone cubic zirconia (CZ).

Faceted zircon showing range of colors.

ZIRCON	
CHEMICAL COMPOSITION	$ZrSiO_4$
COLOR	LIGHT BROWN TO REDDISH BROWN, COLORLESS, GRAY, YELLOW, GREEN, BLUE
REFRACTIVE INDEX	1.93–1.98
RELATIVE DENSITY	4.6–4.7
HARDNESS	7.5
CRYSTAL GROUP	TETRAGONAL
CLEAVAGE	INDISTINCT
FRACTURE	CONCHOIDAL
TENACITY	VERY BRITTLE
LUSTER	VITREOUS TO ADAMANTINE
TRANSPARENCY	TRANSPARENT TO TRANSLUCENT, OCCASIONALLY NEARLY OPAQUE
DISPERSION	STRONG, .039
BIREFRINGENCE	.058
PLEOCHROISM	WEAK EXCEPT IN BLUE STONES

LOCATION Australia, Brazil, Burma (Myanmar), Cambodia, Canada, France, Germany, Italy, Nigeria, Norway, Russia, Sri Lanka, Tanzania, Thailand, United States, Vietnam.

DUMORTIERITE

Cabochon

Cameo

Polished

Because it has just one distinct cleavage and an uneven fracture, and lacks the clear transparency required in a gemstone, dumortierite is not faceted. Instead it is made into cabochons or carved into sculptures to bring out its bright greenish-blue color and its vitreous luster. Dumortierite can be mistaken for SODALITE, LAZULITE, or lapis lazuli. However, the last two differ from dumortierite in not being fibrous, and sodalite shows more extensive light-colored areas. Though the most sought-after specimens are violet to blue in color, dumortierite also occurs in white, brown, black, and red colors. It has a bluish-white streak. The mineral is mostly found in the form of fibrous masses but can occur as prismatic columns. It often forms fibrous inclusions in QUARTZ ("dumortierite quartz"), lending it a pale blue color. The minerals associated with dumortierite are numerous, and include quartz, cordierite, KYANITE, ANDALUSITE, and SILLIMANITE.

▲ *Cabochon-cut dumortierite gem.*

The mineral was identified in 1881 by the French mineralogist M. F. Gonnard, who named it for the French palaeontologist Eugène Dumortier (1803–73). Dumortierite is relatively hard, and is employed in industry for making ceramics – as used in spark plugs, for example.

DUMORTIERITE

Its type locality is Beaunan, Rhône-Alpes, eastern France. Other important sources include Ambositra, Madagascar; Altai, southwest Siberia, Russia; Koralpe Mountains, Styria, southeast Austria. In China, dumortierite has been used as a substitute for lapis lazuli. Gem-quality specimens from Nevada, like some from India, are impregnated with quartz. Other US locations include Oreana, Nevada; King County, Washington; and Fremont County, Colorado.

DUMORTIERITE	
CHEMICAL COMPOSITION	$Al_7BO_3(SiO_4)_3O_3$
COLOR	BRIGHT GREENISH BLUE
REFRACTIVE INDEX	1.686–1.723
RELATIVE DENSITY	3.26–3.41
HARDNESS	7–8.5
CRYSTAL GROUP	ORTHORHOMBIC
CLEAVAGE	ONE DISTINCT
FRACTURE	UNEVEN TO HACKLY
LUSTER	VITREOUS
TRANSPARENCY	TRANSPARENT TO TRANSLUCENT
DISPERSION	STRONG, 1.7
BIREFRINGENCE	.014–.027
PLEOCHROISM	STRONG

LOCATION Austria, China, France, Germany, India, Madagascar, New Zealand, Russia, United States.

◄ *Rough specimen of dumortierite.*

The gemstone family "garnet" contains more than 10 different gemstones with a similar chemical structure. The main differences in physical properties of the garnet group are slight variations in color, density, and refractive index. The name garnet comes from the Latin *granatus*, meaning "seed," because it often resembles small round seeds when found.

Garnets are isomorphous, meaning that they share the same crystal structure. This leads to similar shapes and properties. Garnets belong to the cubic crystal class, which produces very symmetrical, cube-based crystals. The most common crystal shape for garnets is the rhombic dodecahedron, a 12-sided crystal with diamond-shaped faces. Although the color red is the one which occurs most frequently, there are also garnets showing different colors of green, pale to bright-yellow, fiery orange, and fine earthy shades. Many of these are rare and beautiful.

Garnets as a group are relatively common in highly metamorphic and some igneous rocks. They form under high temperatures and pressure. Garnets can be used by geologists as an indication of the temperatures and pressures at which the rock was formed.

The jewelry trade tends to use six main garnet types: pyrope, almandine, spessartine, grossular garnet, andradite, and uvarovite. Best known among the garnets are the deep-red almandine garnet and pyrope garnets. The almandine is what most people think of when garnet is mentioned. It is dark and slightly brown or red. The pyrope garnet tends to have less brown in it. Fine-quality pyrope garnet may be confused with a dark ruby, but medium-quality pyrope looks much like almandine. In the late 1960s, a new garnet was discovered which made green an important garnet color. This is the tsavorite, named after the Tsavo area of southeast Kenya. The increasing scarcity of fine emerald has contributed to tsavorite's importance. The very rare demantoid variety of andradite is an emerald green with diamond-like fire. Uvarovite garnet may also be emerald green, but it is found only as a very small stone.

Garnets occur in relative abundance and enjoy widespread use. Today, garnets come mainly from African countries, but are also found in India, Russia, and Central and South America.

▲ *The centerpiece of this Nardi brooch is a carved garnet plaque with a foliate motif. A modified rose-cut garnet sits above the figure's rose-cut pink sapphire turban.*

◄ *The garnet family contains more than 10 different gems. Red varieties are the most common, while the more expensive varieties tend to be green or yellow. This photograph of crystals, faceted garnets, and beads shows the color range of garnets.*

Brilliant

Mixed

Pyrope is the only garnet that is always a shade of red. It is one of the most popular garnets in jewelry. Its name comes from the Greek word *pyropos*, meaning "fiery-eyed."

Pyrope is a magnesium aluminum silicate and occurs mainly as rounded grains. Unlike the other garnets, pyrope is not commonly found in metamorphic rock. Most pyrope comes from igneous rocks in varying proportions of ALMANDINE-pyrope mixes. A mixture of two pyrope to one almandine is known as **rhodolite** – a purplish-red gem popular in jewelry. Pyrope is usually clearer and less flawed than almandine. Glass imitations are warmer to the touch than genuine pyrope.

▲ *Step-cut pyrope garnet.*

◀ *Faceted pyrope-garnet gem (left) and faceted rhodolite gem (right).*

PYROPE	
CHEMICAL COMPOSITION	$Mg_3Al_2(SiO_4)_3$
COLOR	DARK RED TO RUBY RED
REFRACTIVE INDEX	1.7–1.73
RELATIVE DENSITY	3.51–3.65
HARDNESS	7.25
CRYSTAL GROUP	ISOMETRIC
CLEAVAGE	ABSENT
FRACTURE	CONCHOIDAL
TENACITY	BRITTLE
LUSTER	VITREOUS
TRANSPARENCY	TRANSPARENT TO TRANSLUCENT
DISPERSION	.022
BIREFRINGENCE	.01
PLEOCHROISM	DISTINCT

LOCATION Czech Republic, Germany, Mauritania, Russia, South Africa, United States.

◀ *Rhodolite crystal fragment.*

PYROPE

In the past, Czechoslovakia provided huge quantities of "Bohemian garnet," popular in the jewelry of the late 19th century. Today, pyrope is still found in Bohemia, Czech Republic (particularly around Bilin), and jewelers continue to use the 19th-century style of tightly packed small pyropes. Pyrope is often found associated with DIAMOND and is common in the Kimberley region of South Africa, where it is known as "Cape ruby." Other sources include the diamond fields of the Yakut region of Siberia, Russia; the Harz Mountains, Lower Saxony, northwest Germany; McBride Province, Queensland, Australia; Monastery Mine, Orange Free State, South Africa; Touajil region of northern Mauritania; Mogok mines of Mandalay, central Burma. In North America, pyrope deposits exist in McKinley County, New Mexico, and Albany County, Wyoming.

▶ *Rounded crystal fragment of pyrope garnet from Mozambique.*

T he most common of the garnets, almandine, sometimes also called the "Ceylon ruby," has been popular in jewelry at least since Roman times. It was given its name by the Roman scholar, Pliny the Elder (23–79), who called it "alabandicus," referring to a type of stone found and worked at Alabanda in Turkey. However, almandine's history goes back even further than that and it has been found in Bronze Age excavations.

Almandine is a member of the pyralspite subgroup of garnets. It is an iron aluminum garnet, and derives its red color from its iron content. While garnets are widely distributed in metamorphic and some igneous rocks, almandine is the most common garnet found in schists and gneisses. Transparent almandines are popular as gemstones, although they tend to be somewhat brittle and the faceted edges are prone to chipping. It is, however, one of the hardest of the garnets and can take heavy wear. The best specimens show a clear and uniform color. It is said to bring spirituality to relationships.

Almandine is usually darker red than PYROPE, but the two are closely related, and most red garnets are a hybrid of the two. Pure almandine and pure pyrope are both very rare. In fact, the make-up of SPESSARTINE, almandine and pyrope runs along a continuum as the manganese content of spessartite gives way to the iron of almandine or the magnesium of pyrope. One way to establish the type of garnet is with a density test.

Cabochon

Mixed

▲ *Red almandine garnets in garnet mica schist.*

ALMANDINE

CHEMICAL COMPOSITION	$Ca_{19}Fe(Mg,Al)_8A_{14}$ $(SiO_4)_{10}(Si_2O7)_4(OH)_{10}$
COLOR	RED
REFRACTIVE INDEX	1.83
RELATIVE DENSITY	4.3
HARDNESS	6.5-7.5
CRYSTAL GROUP	ISOMETRIC
CLEAVAGE	NONE
FRACTURE	CONCHOIDAL
TENACITY	BRITTLE
LUSTER	VITREOUS
TRANSPARENCY	TRANSPARENT TO TRANSLUCENT
DISPERSION	MODERATELY HIGH,.024–.027
BIREFRINGENCE	NONE
PLEOCHROISM	NONE

LOCATION India, Italy, Norway, Pakistan, Russia, Sweden, United States, Zimbabwe.

ALMANDINE

Sources of almandine include Nordland, Norway; Northwest Frontier Province, Pakistan; Rajasthan, India; and Kola Peninsula, Russia. Notable US locations include Wrangel, Alaska; Middlesex County, Connecticut; and Oxford County and Androscoggin County, Maine.

▶ *Brilliant-cut almandine gem from Sri Lanka.*

Brilliant

Step

Cabochon

SPESSARTINE

Gem-quality stones were extremely rare until the 1991 discovery of a vein at Kombat Mine, Otavi, northern Namibia. Since then, beautiful orange stones with few inclusions have also been mined in Nigeria and Fujian, China, but sources remain limited. The gemstones should be faceted to bring out the best color, and they benefit from high refraction. Specimens larger than 10 carats are very unusual. Other locations include Nuristan, Afghanistan; Sahatany Valley, Madagascar; and Baltistan, Pakistan. Major US sources include San Diego County, California, and Middlesex County, Connecticut.

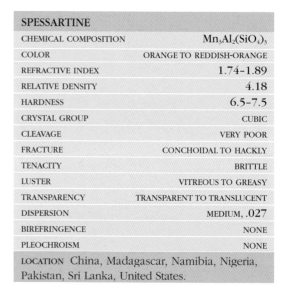

Also known as Mandarin garnet for its vivid orange color, spessartine is one of the rarest members of the garnet family. It is the manganese-rich end member of the pyralspite group. As it approaches ALMANDINE in its chemical composition, it becomes a darker red, and as it moves toward ANDRADITE, it becomes almost black. But the pure orange stone is the most sought after of the spessartine garnets.

▲ *Pendeloque-cut spessartine gem.*

Spessartine is found in manganese-rich metamorphic environments, as well as in some granite pegmatites. It is named for the region of Spessart, Bavaria, southern Germany.

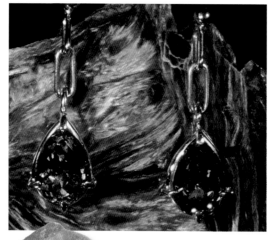

▲ *Pair of spessartine garnet earrings.*

◀ *Spessartine crystal fragment.*

SPESSARTINE	
CHEMICAL COMPOSITION	$Mn_3Al_2(SiO_4)_3$
COLOR	ORANGE TO REDDISH-ORANGE
REFRACTIVE INDEX	1.74–1.89
RELATIVE DENSITY	4.18
HARDNESS	6.5–7.5
CRYSTAL GROUP	CUBIC
CLEAVAGE	VERY POOR
FRACTURE	CONCHOIDAL TO HACKLY
TENACITY	BRITTLE
LUSTER	VITREOUS TO GREASY
TRANSPARENCY	TRANSPARENT TO TRANSLUCENT
DISPERSION	MEDIUM, .027
BIREFRINGENCE	NONE
PLEOCHROISM	NONE
LOCATION	China, Madagascar, Namibia, Nigeria, Pakistan, Sri Lanka, United States.

The orange version of GROSSULAR GARNET, hessonite is also sometimes known as "cinnamon stone" because of its distinctive brownish-orange color.

Like other grossular garnets, hessonite is usually found in metamorphically impure limestones. Hessonite derives its orange color from iron and manganese content.

Hessonite has been a popular stone for thousands of years. The ancient Greeks and Romans used the stones in jewelry, faceting it and setting it in silver or gold, or using it in cameos and intaglio work. It gets its name from the ancient Greek word for "inferior," *esson*, because it is noticeably less hard and less dense than other varieites of garnet. In the past, many hessonite gems were mistakenly identified as the orange variety of ZIRCON (sometimes known as hyacinth). The difference is readily detected by measuring their relative densities: hessonite has a relative density between 3.64 and 3.69, whereas zircon has a relative density of approximately 4.6.

Hessonite is also an important stone in Vedic astrology. Known in Hindu as *gomedha*, it is believed to have been formed from the fingernails of the great demon Vala, which were scattered among the lakes of the East. When it is set in gold, it is believed to be a powerful talisman, increasing happiness and lifespan.

Most hessonite is rich in inclusions, which give it a swirly, treacle-like appearance.

Brilliant

Mixed

Cushion

▲ *Cushion-cut hessonite garnet.*

HESSONITE	
CHEMICAL COMPOSITION	$Ca_3Al_2Si_3O_{12}$
COLOR	ORANGE TO BROWNISH-ORANGE
REFRACTIVE INDEX	1.73–1.75
RELATIVE DENSITY	3.64–3.69
HARDNESS	6.5–7.0
CRYSTAL GROUP	ISOMETRIC
CLEAVAGE	NONE
FRACTURE	CONCHOIDAL TO HACKLY
TENACITY	BRITTLE
LUSTER	VITREOUS TO GREASY
TRANSPARENCY	TRANSPARENT TO TRANSLUCENT
DISPERSION	MEDIUM, .027
BIREFRINGENCE	NONE
PLEOCHROISM	NONE
LOCATION Canada, Czech Republic, Germany, Italy, Russia, Sri Lanka.	

HESSONITE

The clearest stones are the most prized. Major sources of hessonite include Asbestos, Québec, Canada; St Barbora Adit, Bohemia, Czech Republic; Bavaria, Germany; Valle d'Aosta, northwest Italy; Chelyabinsk Oblast, Urals, central Russia; Zermatt in the Alpine canton of Valais, southern Switzerland; and Okkampitiya, southeast Sri Lanka. Hessonite deposits in the United States include Redding, Fairfield County, Connecticut, and Middlesex County, Massachusetts.

▲ *Hessonite crystals on matrix.*

Brilliant

Bead

Polished

Although in its pure state, grossular garnet is clear and colorless, it often takes on color from impurities it contains. For this reason, the stone can be found in a wide range of colors, and some of its varieties are popular as gemstones. It derives its name from the Greek word *grossularia*, for "gooseberry," because its crystals sometimes resemble the fruit in both color and shape.

Some colored grossular is found in its massive state, such as the pink grossular that derives its color from the presence of iron. The massive green grossular known as **Transvaal jade** resembles JADEITE and sometimes contains black specks of magnetite. However, the transparent gemstone varieties of grossular, like orange HESSONITE, are more popular and valuable.

The most sought-after gem variety of grossular is known as **tsavorite** (or tsavolite). Tsavorite relies on the same trace impurities for its color as emerald – vanadium and sometimes chromium. However, its higher refractive index and double dispersion give it a much brighter sparkle. In terms of color, the best specimens of tsavorite are easily comparable to emeralds, and should be selected for their intense, medium-green color, avoiding any with a yellowish tinge. Because the color is similar, tsavorite is often found in an emerald cut, but in fact it displays its sparkle best when cut as a brilliant.

▶ *Square-cut pink grossular garnet.*

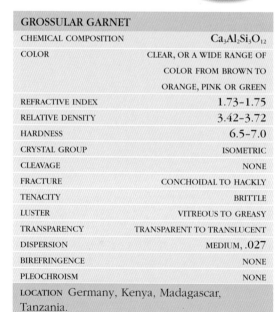

▶ *Oval brilliant-cut tsavorite garnet.*

GROSSULAR GARNET

This vivid green stone was discovered in 1968, in the Merelani Hills, near Arusha, northeast Tanzania. In 1971, a further vein was found at Voi, Coast Province, southeast Kenya. It was named by Tiffany and Co., the first jeweler to market the gemstone, after the Tsavo National Park in southeast Kenya.

Virtually all tsavorite is found in East Africa, in a region rich in vanadium, the trace element that endows the gem with its green color. Found in graphitic gneisses, deposits of tsavorite are small and unpredictable, but new veins are constantly being uncovered. In 1991, the gem was found in Madagascar.

▶ *Gooseberry-green colored crystals of grossular garnet.*

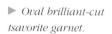

GROSSULAR GARNET	
CHEMICAL COMPOSITION	$Ca_3Al_2Si_3O_{12}$
COLOR	CLEAR, OR A WIDE RANGE OF COLOR FROM BROWN TO ORANGE, PINK OR GREEN
REFRACTIVE INDEX	1.73–1.75
RELATIVE DENSITY	3.42–3.72
HARDNESS	6.5–7.0
CRYSTAL GROUP	ISOMETRIC
CLEAVAGE	NONE
FRACTURE	CONCHOIDAL TO HACKLY
TENACITY	BRITTLE
LUSTER	VITREOUS TO GREASY
TRANSPARENCY	TRANSPARENT TO TRANSLUCENT
DISPERSION	MEDIUM, .027
BIREFRINGENCE	NONE
PLEOCHROISM	NONE
LOCATION Germany, Kenya, Madagascar, Tanzania.	

A fairly common mineral, andradite is found in a wide range of colors. It is popular with collectors as a specimen stone, both rare and valuable, but DEMANTOID, its emerald-green variety, is a highly sought-after gemstone.

Andradite forms in contact or regional metamorphic environments, especially marbles or skarns. The most famous andradite deposit is found in the Ural Mountains, central Russia, but this source is now largely depleted. The stone is named after the famous Brazilian mineralogist J. B. de Andrada e Silva (1763–1838), who first described it.

Brilliant

Mixed

▲ *Black crystals of melanite garnet on matrix.*

ANDRADITE

Andradite gets its color from impurities of manganese, titanium, aluminum, or chromium. The different colors of andradite have their own names: the yellow version, found largely in Switzerland and Italy, is **topazolite**, while the black version, containing titanium, is **melanite**. Melanite is sometimes found in volcanic lava. In general, large, clean stones are rare. Andradite's type localities are Feiringen and Drammen, southern Norway.

ANDRADITE	
CHEMICAL COMPOSITION	$Ca_3Fe_2[SiO_4]_3$
COLOR	GREEN, YELLOWISH GREEN, BROWN, BLACK
REFRACTIVE INDEX	1.85–1.89
RELATIVE DENSITY	3.7–4.1
HARDNESS	6.5
CRYSTAL GROUP	CUBIC
CLEAVAGE	NONE
FRACTURE	CONCHOIDAL
TENACITY	BRITTLE
LUSTER	VITREOUS TO ADAMANTINE
TRANSPARENCY	TRANSPARENT TO TRANSLUCENT
DISPERSION	MEDIUM, .027
BIREFRINGENCE	NONE
PLEOCHROISM	NONE
LOCATION Italy, Mexico, Norway, Russia, Switzerland, United States.	

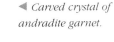

◀ *Carved crystal of andradite garnet.*

▼ *Polished crystal of andradite garnet.*

◀ *Green andradite crystals and brilliant-cut gemstone.*

Brilliant

Mixed

O ne of the rarest and most valuable of gemstones, demantoid is the green variety of ANDRADITE garnet. Its name comes from the Dutch *demant*, meaning "diamond," because its fire is greater even than diamond, and at its best, its color rivals the EMERALD.

Demantoid was first discovered in the Ural Mountains of central Russia in 1868. The Ural stones were distinguished by "horsetail" inclusions of byssolite fibers. They soon became popular, not only with jewelers like Carl Fabergé and Tiffany's of New York, but also among the most discerning collectors of the Victorian era. The Russian tsars were known to be particularly fond of jewelry set with demantoids.

Demantoid is usually fashioned as brilliants or cushion cuts. Stones with lighter color have greater fire, and so the buyer has to choose between body color or fire. Because it is soft, it is not well suited to use in rings. Specimens are generally small, around 1 carat.

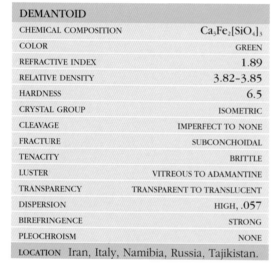

DEMANTOID	
CHEMICAL COMPOSITION	$Ca_3Fe_2[SiO_4]_3$
COLOR	GREEN
REFRACTIVE INDEX	1.89
RELATIVE DENSITY	3.82–3.85
HARDNESS	6.5
CRYSTAL GROUP	ISOMETRIC
CLEAVAGE	IMPERFECT TO NONE
FRACTURE	SUBCONCHOIDAL
TENACITY	BRITTLE
LUSTER	VITREOUS TO ADAMANTINE
TRANSPARENCY	TRANSPARENT TO TRANSLUCENT
DISPERSION	HIGH, .057
BIREFRINGENCE	STRONG
PLEOCHROISM	NONE
LOCATION Iran, Italy, Namibia, Russia, Tajikistan.	

▲ *The outer row of this brooch (c. 1870) features oval demantoid garnets with rose-cut diamond points.*

▶ *Crystal fragment of demantoid garnet.*

▶ *Faceted demantoid garnets.*

DEMANTOID

Demantoid garnets have always been very rare and, after World War I, the gem's scarcity forced it into obscurity. However, the discovery of a new source of the stone, in Namibia in the 1990s, reminded collectors of its appealing qualities. The Namibian stones have excellent color, ranging from light green to a bluish green. However, the horsetail inclusions that characterize the Russian stones are absent, and as a result the Namibian stones are less valuable – a rare example of a gem that increases its value because of an inclusion. At the same time, a number of mines in the Urals, once considered exhausted, have returned to production, although on a small scale.

With its beautiful emerald-green color, uvarovite is the only member of the garnet family to be consistently green. But despite its striking color, uvarovite is rarely used as a gem because its crystals are normally too small to be cut.

Uvarovite is formed from the metamorphism of impure siliceous limestones and some other chromium-containing rocks. It is this chromium that gives uvarovite its characteristic color. It is named for Count Uvarov (1765–1855), a Russian statesman who made it his life's work to catalogue the mineral wealth of his country. Uvarov was the first to write about this member of the garnet family.

Uvarovite is sometimes found as a druse – fine crystal clusters on the host matrix. In this form, it is too fragile to be used for rings but is suitable for other types of jewelry.

Brilliant

UVAROVITE

Some of the best specimens are from localities in Russia and California; others have been found at the Mokkivaara copper mine, Outokumpu, Finland. *Outokumpu* means "hill of the devil" or "strange hill." A small trench has been cut at the edge of the open-cast copper mine for use by tourists and mineral collectors keen to search for uvarovite.

UVAROVITE	
CHEMICAL COMPOSITION	$Ca_3Cr_2[SiO_4]_3$
COLOR	GREEN
REFRACTIVE INDEX	1.86
RELATIVE DENSITY	3.4–3.8
HARDNESS	6.5–7.0
CRYSTAL GROUP	CUBIC
CLEAVAGE	NONE
FRACTURE	CONCHOIDAL
TENACITY	BRITTLE
LUSTER	VITREOUS
TRANSPARENCY	TRANSPARENT TO TRANSLUCENT
DISPERSION	——
BIREFRINGENCE	NONE TO .0005
PLEOCHROISM	NONE
LOCATION	Finland, Italy, Russia, Turkey, South Africa, United States.

▲ *Crystal fragment of uvarovite garnet.*

◀ *Uvarovite garnet crystals on matrix.*

ANDALUSITE

Brilliant

Baguette

An aluminum silicate, andalusite was discovered at Almeria in the southern Spanish province of Andalusia, from where it gets its name. It is heated to form mullite, a refractory material with industrial uses such as spark plugs.

Andalusite is a polymorph with two other minerals, the triclinic KYANITE and the fibrous orthorhombic SILLIMANITE. Some crystals have carbonaceous inclusions, arranged so that in cross-section they form a dark cross. This variety is called **chiastolite**, named after the Greek letter *chi* (symbol Χ or "cross"). Chiastolite existed in schists near the town of Santiago de Compostela, northwest Spain, and many amulets of the "cross stone" were sold to pilgrims. Other important sites of chiastolite include: Bimbowrie, Australia; Hunan Province, China; Siberia, Russia; Brittany, France; Cumbria, England; Massachusetts, United States.

▲ *Cushion-cut andalusite gem.*

Andalusite occurs typically in thermally metamorphosed pelitic rocks, and in pelites that have been regionally metamorphosed under low-pressure conditions. It also occurs, together with CORUNDUM, TOURMALINE, TOPAZ, and other minerals, in some pegmatites (for example, Velké Meziříčí, Moravia, Czech Republic).

Transparent green andalusite is a gemstone of top quality. Unlike other pleochroic gemstones, such as IOLITE and ZOISITE, where gem-cutters try to reduce the pleochroism and highlight the single best color, cutters of andalusite attempt to get a good mix of colors in the stone.

ANDALUSITE

Dark-green andalusite gemstones come from alluvial gravels in Sri Lanka, while green andalusite gemstones are found in Brazil, principally the states of Espírito Santo and Minas Gerais. Brazilian andalusites are small, rare, and very expensive.

Large, columnar crystals of andalusite occur at Lisenz, Austria. Major US locations of andalusite include the White Mountains, California; Standish, Maine; and Delaware County, Pennyslvania.

◀ *Andalusite crystals on matrix.*

▼ *Andalusite (chiastolite) crystal.*

ANDALUSITE	
CHEMICAL COMPOSITION	Al_2SiO_5
COLOR	BLUE, GREEN, BROWN
REFRACTIVE INDEX	1.71–1.73
RELATIVE DENSITY	3.56–3.68
HARDNESS	4.5–7.0
CRYSTAL GROUP	TRICLINIC
CLEAVAGE	DISTINCT
FRACTURE	UNEVEN
TENACITY	BRITTLE
LUSTER	VITREOUS, PEARLY ON CLEAVAGE PLANE
TRANSPARENCY	TRANSPARENT TO TRANSLUCENT
DISPERSION	MEDIUM, .02
BIREFRINGENCE	.01
PLEOCHROISM	DISTINCT
LOCATION Austria, Australia, Brazil, China, Czech Republic, France, India, Russia, Spain, Sri Lanka, United Kingdom, United States.	

Beautiful and rare, sillimanite is found in two forms: one clear and glassy, and the other silky and fibrous. It is named for the US geologist Benjamin Silliman (1779–1864). When it is found in its fibrous state, it is often known as **fibrolite**.

Sillimanite occurs typically in schists and gneisses produced by high-grade regional metamorphism. It is generally scattered among layers of metamorphic rocks, often around hot springs or areas of volcanic activity.

While mineral collectors prize the fibrous crystals, the glassy crystals are used as gemstones. However, gem-quality stones are very rare, and they are also brittle and difficult to cut. Some crystals demonstrate a cat's-eye effect, and when they are polished as cabochons they glow mysteriously, like a cat's eyes at night. These cat's-eye cabochons make beautiful rings. When fibrolite is polished, the straight, parallel crystals are clearly visible.

Cushion

Cabochon

SILLIMANITE

CHEMICAL COMPOSITION	Al_2SiO_5
COLOR	COLORLESS TO WHITE, YELLOW, BLUE OR BROWN
REFRACTIVE INDEX	1.65–1.68
RELATIVE DENSITY	3.24
HARDNESS	6–7
CRYSTAL GROUP	ORTHORHOMBIC
CLEAVAGE	PINACHOIDAL, GOOD
FRACTURE	UNEVEN
TENACITY	BRITTLE
LUSTER	VITREOUS
TRANSPARENCY	TRANSPARENT TO TRANSLUCENT
DISPERSION	.015
BIREFRINGENCE	.018–.022
PLEOCHROISM	DISTINCT

LOCATION Burma (Myanmar), Czech Republic, Sri Lanka, United States.

SILLIMANITE

The clear, blue stones, found in Ohngaing, Mogok, Burma (Myanmar), are the most sought after. Gray to green stones come from Sri Lanka. The larger stones found in the Clearwater River Valley in Idaho, USA, are often made into carvings. Type localities are Moldau and Schuttenhoten, Czech Republic, and Chester, Middlesex County, Connecticut, USA.

◀ *Blue cushion-cut sillimanite gem.*

▼ *Sillimanite cat's-eye cabochon.*

◀ *Sillimanite crystals on matrix.*

KYANITE

Step

Baguette

Cabochon

K yanite is an attractive mineral in-frequently used as a gemstone in jewelry. Its brittleness means that extreme care must be taken when handling, cutting, and polishing the stone. Kyanite has perfect cleavage in one direction. Its name derives from the Greek *kyanos*, meaning "blue." It is used in the manu-facture of spark plugs.

Kyanite is unique among gems in that it has a wide variation in hardness across the same crystal – being 4.5 along the length and 7 across the width. It is also known as disthene because of this variation in hardness.

Kyanite occurs in metamorphic rocks such as schist and gneiss. Associated minerals include andalusite, corundum, and staurolite.

In jewelry, the most prized kyanites are transparent and deep cornflower blue, pale blue, or bluish-green in color. Kyanite can be mistaken for sapphire or aqua-marine. It is often fashioned into long rectangular and step cuts or into oval cabochons. It sometimes has magnetic properties similar to tourmaline.

▲ *Step-cut kyanite gem.*

KYANITE

Kyanite is extracted from alluvial deposits at Villa Rica, Brazil. Rare, color-less kyanite is found at Machakos, Kenya. The schists of St Gotthard in the Tyrol Mountains of Switzerland yield very fine kyanite crystals in association with staurolite. Major US locations include Yancey County, North Carolina, and Madison County, Montana.

KYANITE	
CHEMICAL COMPOSITION	Al_2SiO_5
COLOR	BLUE, GREEN, BROWN
REFRACTIVE INDEX	1.71–1.73
RELATIVE DENSITY	3.56–3.68
HARDNESS	4.5–7.0
CRYSTAL GROUP	TRICLINIC
CLEAVAGE	DISTINCT
FRACTURE	UNEVEN
TENACITY	BRITTLE
LUSTER	VITREOUS, PEARLY ON CLEAVAGE PLANE
TRANSPARENCY	TRANSPARENT TO TRANSLUCENT
DISPERSION	MEDIUM, .02
BIREFRINGENCE	.01
PLEOCHROISM	DISTINCT

LOCATION Brazil, China, India, Kenya, Mozambique, Nepal, Russia, Serbia, Switzerland, United States.

◀ *Kyanite crystals in matrix.*

Often known as "fairy stones" or "fairy crosses," staurolite intrigues collectors because it occurs naturally in the shape of a cross. Named for the Greek *stauros* and *lithos*, or "cross stone," for centuries it has been prized as a good-luck talisman and worn as an amulet. Today, genuine specimens are rare, and many staurolites on the market are actually other stones carved into the sought-after shape. Real staurolite can be determined by its ability to scratch glass.

Staurolite displays the classic penetration form of twinned crystals. It forms two types of twins, one at 60 degrees (known as St Andrew's crosses) and one at 90 degrees (known as Greek crosses). Rare specimens display both types of twinning and resemble roughly drawn stars.

Many legends are attached to this gem. One says that the crosses are formed by the tears of fairies falling upon the ground. Another says that they are the tears of the Cherokee tribe of Native North Americans, following the "Trail of Tears" (1838) to reservations after they were forced to leave their ancestral lands. This tragic journey reduced the Cherokee population by more than 25%. In addition to cross-shaped specimens, serious collectors look out for transparent, faceted staurolites. These are extremely rare, because it is unusual to find a staurolite crystal that is both light-colored and large enough to be faceted.

Step

Baguette

Cameo

◄ *Staurolite crystals on matrix.*

STAUROLITE	
CHEMICAL COMPOSITION	$(Fe_2+,Mg,Zn)_2Al_9$ $(Si,Al)_4O_{22}(OH)_2$
COLOR	REDDISH-BROWN TO BROWN-BLACK
REFRACTIVE INDEX	1.739–1.792
RELATIVE DENSITY	3.65–3.77
HARDNESS	7–7.5
CRYSTAL GROUP	MONOCLINIC, PSEUDO-ORTHORHOMBIC
CLEAVAGE	ONE, DISTINCT CLEAVAGE
FRACTURE	SUBCONCHOIDAL
TENACITY	———
LUSTER	VITREOUS TO RESINOUS
TRANSPARENCY	TRANSLUCENT TO NEARLY OPAQUE
DISPERSION	.023
BIREFRINGENCE	.011–.015
PLEOCHROISM	DISTINCT TRICHROIC
LOCATION Brazil, England, France, Switzerland, United States.	

STAUROLITE

Staurolite occurs typically as prophyro-blasts in medium-grade schists and gneisses, often in association with GARNET, KYANITE, and mica. Cherokee County, North Carolina, USA, remains a major source of the gem.

▶ *Staurolite crystals occur in the shape of a cross.*

Topaz is a hydrous aluminum fluorosilicate that occurs typically in the veins and cavities of granite pegmatites, rhyolites, and quartz veins. The hardest silicate mineral and one of the hardest minerals, topaz makes excellent mineral specimens because of its high luster, attractive colors, and well-formed crystals. It is found in granites that have been subjected to alteration by fluorine-bearing solutions, and can be accompanied by FLUORITE, TOURMALINE, APATITE, BERYL, QUARTZ, and the tin-bearing mineral CASSITERITE. Topaz can thus be an indicator of the presence of tin ore. It also occurs as worn pebbles in alluvial deposits. Perfect cleavage helps identify topaz on sight from similar minerals.

▲ *Pale green step-cut topaz.*

The name comes from the Greek *topazion*, which may originate from the Sanskrit *tapas*, meaning "fire," or from the Egyptian island of Topazos (now Zebirget or St John's Island) in the Red Sea. The Latin writer Pliny the Elder (AD 23–79) used the island's name for a yellowish-green stone found there, and it soon became the name for most yellow stones.

▲ *Various colors and cuts of topaz gems.*

▶ *Yellow topaz crystal.*

TOPAZ

The largest topaz crystals are found at Minas Gerais, Brazil, where specimens with a width of 12 inches (30 centimeters) have been discovered. They have a sherry-brown color. Brazil also yields honey-yellow, red, pink, blue, and white crystals, and topaz pebbles known as *pingos d'água* or "drops of water." Blue, yellow, pink, and colorless stones occur in the Urals in Russia, and colorless and pale brown varieties are found in Queensland and Tasmania, Australia. The United States has colorless, yellow, reddish, and pale blue crystals, with sources in San Diego County, California; Juab County, Utah; the Pike's Peak area of Colorado; Lords Hill in Maine; and in other states, including New Hampshire, Connecticut, Virginia, Texas, and Utah.

Other deposits include dark wine-yellow stones from Siberia in Russia, and Mino province in Japan. There are occurrences also in Durango in Mexico, and Australia, Ireland, Norway, Sweden, Germany, Russia, Ukraine, Pakistan, Nigeria, and Sri Lanka. Topaz that is not of gem quality is found in Northern Ireland, the Cairngorm Mountains of Scotland, and in Cornish tin mines, as well as St Michael's Mount and Lundy Island off the Cornwall and Devon coasts of England.

Brilliant

Step

Pendeloque

◀ *Three-strand blue topaz necklace. Each strand is made up of faceted blue topaz beads.*

Cushion

Mixed

Topaz produces some of the largest crystals, weighing up to 660 pounds (300 kilograms). The largest cut topaz, the pale blue "Brazilian Princess" found at Teofilo Otoni north of Rio de Janeiro, weighs 21,327 carats and was fashioned as a square cut. A smaller but even better-known stone is the "Braganza" from Ouro Preto, Minas Gerais, Brazil, which was mistaken for a diamond when discovered in 1740. It was set in the Portuguese royal crown.

Crystals are normally attached at one end to rock cavities and will break along a flat surface when detached. The columnar crystals of topaz are usually prismatic, often with two or more vertical prism forms and with vertical striations on the faces. Massive, granular forms also exist. The crystals occur in a variety of colors, according to the amount of metallic trace elements present when formed, such as iron and chromium. Dark-orange topaz is called **hyacinth**. The less valuable CITRINE is sometimes sold as "Brazilian topaz."

TOPAZ	
CHEMICAL COMPOSITION	$Al_2SiO_4(OH,F)_2$
COLOR	COLORLESS, PALE YELLOW, PALE BLUE, GREENISH, PINK
REFRACTIVE INDEX	1.60–1.63
RELATIVE DENSITY	3.5–3.6
HARDNESS	8
CRYSTAL GROUP	ORTHORHOMBIC
CLEAVAGE	PERFECT
FRACTURE	SUBCONCHOIDAL TO UNEVEN
TENACITY	BRITTLE
LUSTER	VITREOUS
TRANSPARENCY	TRANSPARENT TO TRANSLUCENT
DISPERSION	WEAK, .014
BIREFRINGENCE	.01
PLEOCHROISM	DISTINCT

LOCATION Australia, Brazil, England, Germany, Ireland, Japan, Mexico, Nigeria, Norway, Pakistan, Russia, Scotland, Sri Lanka, Sweden, Ukraine, United States.

▶ *Navette-cut Belle Epoque (c. 1910) pink topaz and diamond pendant. The cushion-cut pink topaz is suspended from a diamond collet.*

The ancient Greeks and Romans greatly valued topaz as a gemstone. Small wine-yellow "Saxonian topaz" was mined at Schneckenstein in the Erzgebirge Mountains, Saxony, Germany, in medieval times, and these specimens were worn in jewelry by several rulers. Deep mining was later used at the site from 1737 to 1800. Topaz was a prized but rare stone from the Middle Ages until discoveries of large deposits in Brazil in the mid-19th century. Now it is a popular and very affordable gem.

A distinctive feature of topaz is its perfect, easy cleavage. This requires careful handling when stones are cut and polished, since specimens may split or develop internal cracks. Even the warmth of a hand can cause cracks within. Colorless topaz is often cut as round or oval brilliants. Pendeloque stones in the mixed-cut style are also popular. Colored topaz is frequently cut as trap and table cuts. The stones will take a very smooth polish.

The clear or pink, blue, and honey-yellow varieties are especially valued. The most sought-after and expensive colors are called "imperial topaz." Pale pink topaz is often backed by red foil to increase its color. Orange stones are moderately priced. Some varieties, such as purplish ones, are treated with heat or radiation to enhance the color, and pink ("burnt") gem topaz is obtained by heat-treating the dark yellow stones. Yellowish-brown specimens from Brazil turn red. Stones with a slight blue color are normally irradiated with gamma rays to darken the color. Extended heating will produce colorless stones, but colorless topaz will turn brown when radiated.

▶ *Selection of different colors and cuts of topaz. These pale-colored topaz stones can be extremely valuable.*

Step

Table

A rare mineral, euclase sometimes has the clarity to be used as a gemstone. It is a silicate of beryllium and aluminum, as is beryl, but euclase also contains hydrogen. Its name comes from the Greek words *eu* ("well") and *klasis* ("breaking"), which refer to its easy cleavage. The first samples were discovered in South America and sent to Europe in 1785. Euclase sometimes occurs in gem-quality crystals, but its brittleness means that they seldom exceed 1 inch (2 centimeters) in length.

Euclase has become popular with mineral collectors, but special care must be taken when cutting and polishing gemstones due to the tendency of the well-formed crystals to chip. This is because the cleavage is parallel to the plane of symmetry. Euclase gems are most often rendered into step and table cuts, with the bluish-green color being the most valued. Some specimens suffer from an uneven color distribution. Once cut, euclase resembles TOPAZ and types of beryl, such as AQUAMARINE. The crystal's fragility also causes problems with the gemstone's durability.

Euclase is found in granite pegmatites in association with other beryllium minerals, especially beryl. It sometimes occurs as small crystals next to FELDSPARS. The crystals are vertically striated prisms that have slanted termination faces with acute pyramids. Twinning is unknown, and foreign impurities are rare. Euclase may be discovered alongside GOLD as eroded pebbles in rivers and streams. Other associated minerals include QUARTZ, micas, and pericline.

EUCLASE

Euclase is found at several locations in the southeastern Brazilian state of Minas Gerais, especially the type locality of Ouro Preto, and at Alto Equador, Rio Grande do Norte, northeast Brazil. Other major sites include Chivor, Boyacá, central Colombia; Fichtelgebirge, Bavaria, southern Germany; Yakutia, northeast Siberia, Russia; Okehampton, Devon, southwest England; and Fairfield County, Connecticut, USA.

Blue euclase from Mwame, Hurungwe District, Mashonaland, northern Zimbabwe, was first collected in the late 1970s.

EUCLASE	
CHEMICAL COMPOSITION	$BeAlSiO_4OH$
COLOR	COLORLESS TO
	PALE BLUE-GREEN
REFRACTIVE INDEX	1.65–1.67
RELATIVE DENSITY	3.0–3.1
HARDNESS	7.5
CRYSTAL GROUP	MONOCLINIC
CLEAVAGE	ONE PERFECT
FRACTURE	CONCHOIDAL
TENACITY	BRITTLE
LUSTER	VITREOUS
TRANSPARENCY	TRANSPARENT TO TRANSLUCENT
DISPERSION	WEAK, .016
BIREFRINGENCE	.02
PLEOCHROISM	DISTINCT

LOCATION Brazil, Colombia, Germany, Mozambique, Russia, Sri Lanka, United Kingdom, United States, Zimbabwe.

▲ *Oval step-cut euclase, 5.97 carats.*

◄ *Euclase crystal.*

EPIDOTE

Cushion

Step

E pidote is at the iron-rich end of a chemical series, at the other end of which lies clinozoisite. Both these minerals are widespread in medium- to low-grade metamorphic rocks, especially those derived from igneous rocks such as basalt and diabase, or from calcareous sediments. They also occur in contact-metamorphosed limestones, and in veins in igneous rocks.

Epidote's color, dark green to black, masks its optical properties. One variety of epidote has a distinctive pistachio-green color, which has led to its being dubbed "pistacite." The lighter green and yellow transparent specimens can be faceted, though they tend to result in smaller gems of less than 5 carats. When polished to form cabochons, epidote can produce cat's-eye stones.

Epidote's distinguishing features are its distinctive yellow-green color and prismatic habit. Epidote can be mistaken for TOURMALINE, but the latter lacks cleavage and has a hexagonal or triangular cross-section.

EPIDOTE

Epidote is widely distributed. Its type locality is Bourg d'Oisans, Isère, eastern France. Fine crystals are found at Krasnoyarsk, central Siberia, Russia. Other European locations include the Jílove district, Bohemia, Czech Republic; Erzgebirge, Saxony, east Germany; Valle d'Aosta, northwest Italy, and Liguria, northern Italy; Nora, Västmanland, eastern Sweden; Isle of Mull, western Scotland; and the Lizard Peninsula, Cornwall, southwest England.

In the Americas, epidote is found in Baja California, Mexico, and a rare colorless variety has been discovered at Tierra del Fuego, Argentina. In the US, fine specimens come from Fairfield County, Connecticut; Middlesex County, Massachusetts; Houghton County, Michigan; Grafton County, New Hampshire; Sussex County, New Jersey; Sterling Lake, Orange County, New York; Berkeley, Rhode Island; Buckhorn Mountain, Washington; and Green Monster Mine, Prince of Wales Island, Alaska. In Canada, major sites include Bancroft District, Hastings County, and Cobalt, Timiskaming County, Ontario. Other locations include Macquarie Island, Tasmania, Australia.

◀ *Epidote crystals.*

◀ *Step-cut epidote gem.*

EPIDOTE	
CHEMICAL COMPOSITION	$Ca_2(Al,Fe_3+)_3Si_3O_{12}OH$
COLOR	YELLOWISH GREEN TO BLACK
REFRACTIVE INDEX	1.71–1.79
RELATIVE DENSITY	3.3–3.6
HARDNESS	6–7
CRYSTAL GROUP	MONOCLINIC
CLEAVAGE	ONE PERFECT
FRACTURE	UNEVEN
TENACITY	BRITTLE
LUSTER	VITREOUS
TRANSPARENCY	TRANSPARENT TO NEARLY OPAQUE
DISPERSION	STRONG, .019–.03
BIREFRINGENCE	.015–.049
PLEOCHROISM	STRONG
LOCATION Argentina, Australia, Canada, France, Italy, Germany, Mexico, Russia, Sweden, United Kingdom, United States.	

Zoisite is a hydrous calcium aluminum silicate mineral. It was named after Austrian scientist Sigmund von Zois (1747–1819), who discovered it in the Saualpe mountains of Carinthia, Austria, which provided the mineral's first name of "saualpite." Zoisite occurs in schists and gneisses and in metasomatic rocks, together with GARNET, VESUVIANITE, and actinolite. It is found occasionally in hydrothermal veins. Normally it forms aggregates that can be lumpy, fibrous, or bladed. Zoisite can be confused by the naked eye with the more common mineral clinozoisite.

Step

Cabochon

Cameo

▲ *Thulite, the pink variety of zoisite.*

Zoisite generally has long, prismatic or tabular crystals that are commonly striated lengthwise. The blue variety called **tanzanite** is a valuable gemstone discovered in 1967 in Tanzania. **Thulite**, a massive pink variety colored by manganese and found in Sor-Trondelag, Norway, is used in jewelry, polished and carved into cabochons and beads, as well as small ornaments. A massive brilliant-green variety, which contains an irregular spread of rubies and sometimes hornblende, is also polished, carved or tumbled to create ornaments, or a colorful stone used in decorative work. A grayish-green variety from California is also popular for jewelry. Zoisite gems are often heat-treated to enhance their color.

ZOISITE

Sources of zoisite include Värmland, Sweden; Grafton County, New Hampshire, USA; Baltistan, northern Pakistan; and North Cape Province, South Africa.

ZOISITE	
CHEMICAL COMPOSITION	$Ca_2Al_3Si_3O_{12}OH$
COLOR	GRAY, SOMETIMES PINK, PALE GREEN, OR BROWN
REFRACTIVE INDEX	1.69–1.70
RELATIVE DENSITY	2.3–3.4
HARDNESS	6
CRYSTAL GROUP	ORTHORHOMBIC
CLEAVAGE	PERFECT
FRACTURE	UNEVEN
TENACITY	BRITTLE
LUSTER	VITREOUS
TRANSPARENCY	TRANSPARENT TO SUBTRANSLUCENT
DISPERSION	RELATIVELY STRONG
BIREFRINGENCE	.01
PLEOCHROISM	STRONG
LOCATION Austria, Norway, Pakistan, South Africa, Sweden, United States.	

► *Zoisite (thulite), rough.*

◄ *Various shapes and cuts of tanzanite gems*

AXINITE

Brilliant

Mixed

A complex calcium and aluminum borosilicate that normally occurs as thin brown crystals. The name derives from the Greek word *axine*, meaning "axe," which describes the thin wedgelike form of its crystals. Axinite is collected as a mineral specimen, for its excellent luster and well-formed flattened crystals make it a popular choice. Axinite crystals are rarely cut as gems because of their low clarity. A gem-quality variety is found at Bourg d'Oisans, Isère, eastern France.

The members of the axinite group, which have the same structure but are different in terms of chemical composition, include the pale blue magnesio-axinite and the yellow-orange manganaxinite, both magnesium rich; the lilac brown ferro-axinite, iron rich and most common member; and the yellow tinzenite, having iron and manganese. Axinite occurs in calcareous rocks that have undergone contact metamorphism and metasomatism, and also in cavities in granites. Its crystals are normally broad with sharp edges, and also massive, lamellar, or granular. Its associated minerals include QUARTZ, CALCITE, ANDRADITE garnet, DIOPSIDE, EPIDOTE, SCHEELITE, and PREHNITE.

AXINITE

As well as the French Alps, large deposits of axinite exist around St Just, Cornwall, southwest England. Other sites include Timiskaming, Ontario, Canada; Sauerland, North Rhine-Westphalia, western Germany, and the Erzgebirge Mountains, eastern Germany; and Tasmania, Australia.

◀ *Wedge-shaped crystals of axinite.*

AXINITE

CHEMICAL COMPOSITION	$(Ca,Mn,Mg,Fe)_3$ $A_{12}BSi_4O_{15}OH$
COLOR	CLOVE-BROWN, YELLOWISH, GRAY, VIOLET TINGE
REFRACTIVE INDEX	1.66–1.70
RELATIVE DENSITY	3.3–3.4
HARDNESS	6.5–7
CRYSTAL GROUP	TRICLINIC
CLEAVAGE	ONE GOOD DIRECTION
FRACTURE	CONCHOIDAL
TENACITY	—
LUSTER	VITREOUS
TRANSPARENCY	TRANSPARENT TO TRANSLUCENT
DISPERSION	NONE
BIREFRINGENCE	.010–.021
PLEOCHROISM	STRONG

LOCATION Australia, Canada, France, Germany, Russia, United Kingdom.

▶ *Pendeloque-cut manganaxinite gem.*

BERYL

In its pure state, beryl is colorless and of average fire and brilliance, and would attract little notice. But impurities in the stone give it a beautiful range of colors that excite the interest of jewelry lovers and gem collectors alike. The presence of chromium and vanadium produces the brilliant green of the EMERALD, one of the most sought-after gemstones. Iron produces the blues of AQUAMARINE and the yellows of HELIODOR, while manganese gives rise to the pink of MORGANITE and the extremely rare RED BERYL. Because there are so many appealing varieties, beryl is sometimes known as the "mother of gemstones."

Beryl most commonly occurs as an accessory mineral in granites, and is usually found in cavities and in granite pegmatites. Beryl crystals in some pegmatites grow to very large sizes – as much as 30 feet (10 meters). It also occurs in mica schists and gneisses in association with PHENAKITE, RUTILE, and CHRYSOBERYL. Most beryl is not of gem quality. Beryl is the major industrial source of beryllium, which is an extremely strong metal used in the nuclear industry and in alloys for aircraft.

Beryl is one of the oldest recorded stones, and it has been used in jewelry and ornaments for thousands of years. A mine in Egypt, for instance, produced the emeralds that enhanced the famed beauty of Cleopatra. It has also long had practical applications. Goshenite, the clear form of beryl, was used for lenses and the first spectacles – the Roman Emperor Nero is said to have used a monocle cut of beryl.

Collectors are attracted not only by the vivid colors of the beryl varieties but also by the striking crystals. Most are combinations of hexagonal prisms and basal pinacoids. They generally vary from just a few fractions of an inch (centimeter) to almost 3 feet (1 meter); some truly gigantic examples have also been found, weighing several tons. Well-formed crystals can be found embedded in a variety of different stones, making attractive displays.

Today, much beryl is heat-treated or irradiated to improve the color. This is especially true of aquamarines, but also applies to MORGANITE and the other varieties.

▲ *Pair of carved green beryl and diamond ear pendants. A carved green beryl flower supports a pearl-shaped beryl gem.*

▶ *Beryl group showing a small crystal on matrix next to a massive crystal. Below are examples of the various colors and cuts of beryl gems.*

EMERALD

CHEMICAL COMPOSITION	$Be_3Al_2Si_6O_{18}$
COLOR	GREEN
REFRACTIVE INDEX	1.57–1.58
RELATIVE DENSITY	2.67–2.78
HARDNESS	7–8
CRYSTAL GROUP	PRISMATIC
CLEAVAGE	BASAL, POOR
FRACTURE	CONCHOIDAL TO UNEVEN
TENACITY	BRITTLE
LUSTER	VITREOUS
TRANSPARENCY	TRANSPARENT TO TRANSLUCENT
DISPERSION	LOW, .014
BIREFRINGENCE	LOW, .005–.009
PLEOCHROISM	WEAK

LOCATION Afghanistan, Australia, Austria, Brazil, Colombia, India, Madagascar, Mozambique, Pakistan, Russia, South Africa, United States.

▲ *Step-cut emerald within brilliant-cut diamond surround.*

Renowned for its incomparable color, the emerald has been prized for millennia. The ancient Egyptians mined emeralds as early as 3000 BC, from a source by the Red Sea that was later known as "Cleopatra's Mines" – the queen of legendary beauty was known to adore the jewel. The gem is also prized in the *Vedas*, the sacred texts of Hinduism, and was worn as a sacred talisman. In the New World, the Aztecs and the Incas both worshipped the stone, and the famed treasure of Montezuma contained many large and beautiful emeralds. The name derives from the ancient Greek word, *smaragdos*, meaning "green gemstone."

The vivid green of the emerald symbolizes spring and rebirth in many traditions. It also has strong ties to love – the ancient Romans dedicated the color green to Venus, the goddess of love. In the Middle Ages, emerald was used to foresee the future, and also to protect against evil spirits.

Like other members of the BERYL group, emeralds typically occur in granite pegmatites, but also significantly in mica schists and limestone deposits. Normally colorless, beryl takes on its emerald-green color from the presence of chromium and sometimes vanadium. However, not all green beryl is emerald – the presence of chromium distinguishes emerald from ordinary green beryl. Emeralds, like other beryls, form hexagonal crystals, which can sometimes be very large: the biggest ever found, discovered in 1969, weighed in at 7,025 carats. Other huge crystals include the "Devonshire Emerald," an uncut crystal of 1,383.95 carats, and the "Patricia Emerald," at 632 carats (the Patricia Emerald was uncovered by dynamiting the vein, and in the process an even larger crystal was said to have been blown up!).

Practically all emeralds possess inclusions, giving them a soft, mossy internal appearance, which is known as their *jardin* (from the French for "garden"). A transparent stone with a pure,

◄ *The uncut Patricia Emerald, 632 carats.*

► *Beryl (emerald) crystal on matrix.*

green color is the ideal, but the *jardin* does not necessarily detract from the stone's value, assuming that its color is good and that the inclusions do not run so deep that they weaken the stone. In fact, sometimes the stones with the best color are also the most included.

Typically, emeralds are fashioned in a step cut – this is so common that the rectangular step cut is familiarly known as the "emerald cut." The many fissures and inclusions in the stone mean that it is quite tricky to cut, making cut stones of more than 2 carats very expensive. There are specialist emerald cutters in both Jaipur, India, and Tel Aviv, Israel, who devote their lives' work to master the stone.

Emeralds are treated with oils and resins during cutting and polishing to fill fissures and produce a better finish. Sometimes hardeners are incorporated to make the finish longer-lasting. As a result, emeralds must be treated with care: they should not be immersed in detergents or cleaned in ultrasonic cleaners, and they should also be professionally re-oiled every two to three years to maintain their best condition. Today, synthetic emeralds are produced that imitate the color of the best gems but without the inclusions that "fog" the stone, yet most collectors maintain that artificial stones lack the depth and complexity that contribute to the natural beauty of a real emerald gem.

Step

Pendeloque

Cabochon

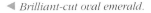

◀ *Brilliant-cut oval emerald.*

▼ *Emerald and diamond brooch with rectangular- and square-cut emeralds.*

EMERALD

The mines of Colombia have historically produced the best-quality emeralds. In the 16th century, the Inca emerald mines were seized by Spanish *conquistadors*, who were as dazzled by the brilliant green stones as they were by the Incan gold. Legend has it that many large and beautiful stones, some reputedly as large as ostrich eggs, went down with ship-wrecks in the Atlantic Ocean, as the *conquistadors* failed to return home with the riches they had plundered. Today, Muzo and Chivor in Boyacá, central Colombia, once worked by the Incas, are still producing lovely gems; those from Chivor tend toward the bluer, cooler end of the spectrum, while the Muzo stones are warmer and more yellow. Today, the Coscuez mines are the largest source of emeralds in Colombia. But while the Colombian mines were once responsible for around 80–90% of world production, today that figure is probably less than 60%. The Copperbelt in northern Zambia, and Bahia, northeast Brazil, are also major sources of top-quality gemstones. North Carolina, USA, has the largest deposits of emerald in North America.

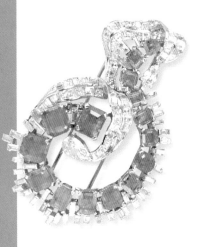

Brilliant

Step

Cabochon

The clear blue of aquamarine attracts jewelry makers and gem collectors alike. Its tone flatters all skin colors and harmonizes with all precious metals. It is relatively easy to cut and can often be found in innovative shapes, as cutters experiment with new forms.

Aquamarine is a member of the beryl group. Beryl commonly occurs as an accessory mineral in granites, and is usually found in cavities and in granite pegmatites. Beryl is usually clear, but iron content gives aquamarine its characteristic blue color. Beryl crystals in some pegmatites grow to very large sizes, even up to 30 feet (10 meters) – aquamarine crystals of up to 3 feet (1 meter) long are not uncommon. Aquamarine is also sometimes found in stream gravels. It is generally free from inclusions, meaning that it is a durable stone; this is also why it is easy to cut. However, inclusions of biotite, RUTILE, PYRITE, and HEMATITE are possible.

Aquamarine is a fairly common stone, which makes it less expensive than the other gems of the beryl group, such as EMERALD. Its name derives from the Latin *aqua*, for water, and *mare*, for sea, and many superstitions and legends about the sea have been attached to the gemstone over the years. It has been said to come from the treasure

▲ *Three brilliant-cut aquamarine gems in a variety of shapes.*

▼ *Pendeloque-cut aquamarine gem.*

▲ *Aquamarine and diamond pendant with large step-cut and rectangular cut-cornered aquamarine.*

▼ *Aquamarine crystals.*

AQUAMARINE

The major source of aquamarine is the state of Minas Gerais, southeast Brazil, especially the mines around Araçuaí. The Brazilian mines produce stones of a characteristic color with distinctive inclusions, and gemmologists can often determine the mine from which a particular stone came based solely on a visual inspection.

In addition to the mines of Minais Gerais, important sources of aquamarine include Kunar Province, Afghanistan; Mianyang City, Sichuan, west-central China; Embu, central Kenya; Zambezia, central Mozambique; Haramosh Mountains of Baltistan, northern Pakistan; and Buryatia Republic, eastern Siberia, Russia. Sites in the United States include Albany and Auburn, Maine.

◄ Huge aquamarine crystal with other beryl crystals of morganite, heliodor, and emerald.

chests of mermaids. Sailors often took aquamarine with them to sea as a lucky charm to protect against shipwreck; sometimes the stone was carved with the image of Poseidon or Neptune, the gods of the sea. When it is immersed in water, the stone is practically invisible and the water in which it was immersed was long considered to have curative properties.

A top-quality aquamarine should be clear and free of inclusions. Today, the most popular stones are a clear blue, avoiding a tinge of either yellow, green, or gray. In the 19th century, jewelry *aficionados* sought out sea-green gems. Because aquamarine's color is usually pale, only larger specimens display any depth of pigment, but smaller stones are lively and decorative, especially when set with diamonds. Aquamarines are generally found in step cuts or as brilliants.

Today, most aquamarine is heat-treated to produce the favored clear, blue color. This treatment reduces ferric-iron content and thus removes any yellow tinge. The treatment is permanent, stable, and acceptable in the market. However, because aquamarine is sensitive to heat, it should be protected from excessive sun and heat exposure.

AQUAMARINE

CHEMICAL COMPOSITION	$Be_3Al_2Si_6O_{18}$
COLOR	BLUE
REFRACTIVE INDEX	1.57–1.59
RELATIVE DENSITY	2.68–2.80
HARDNESS	7.5–8
CRYSTAL GROUP	HEXAGONAL
CLEAVAGE	ONE DIRECTION, POOR
FRACTURE	CONCHOIDAL
TENACITY	BRITTLE
LUSTER	VITREOUS
TRANSPARENCY	TRANSPARENT TO TRANSLUCENT
DISPERSION	LOW, .014
BIREFRINGENCE	LOW, .004–.008
PLEOCHROISM	WEAK

LOCATION Afghanistan, Brazil, China, India, Kenya, Madagascar, Nigeria, Pakistan, Russia.

Baguette

Marquise

Table

T his golden-yellow stone was first discovered at Rössing, Erongo, western Namibia in 1910. Because of its unusually vibrant yellow color, it was named heliodor, from the Greek *helios* and *doron*, meaning "gift from the sun." Initially, only golden beryl found in Namibia was called heliodor, but today the name is applied almost indiscriminately to all varieties of yellow and golden beryl. Strictly, it should only be used for the yellow varieties that get their color from trace iron impurities.

▲ *Octagonal, step-cut heliodor gem.*

Like the other members of the beryl group of gems, heliodor most commonly occurs as an accessory mineral in granites, and is usually found in cavities and in granite pegmatites. Beryl crystals in some pegmatites grow to very large sizes, even up to 30 feet (10 meters), and heliodor is famous for its large, perfect golden crystals. It is sometimes associated with quartz feldspars and muscovite mica.

Heliodor is usually cut as step cuts to enhance its depth of color. When cut as a cabochon, it sometimes displays chatoyancy or asterism. Larger crystals can be carved into ornaments. Its warm, yellow color is said to enhance intuition and compassion.

Like other beryls, heliodor can be oiled or heat-treated. Heating can eliminate the yellow tint, resulting in a clear or even aquamarine blue stone. Irradiation can reverse this effect.

HELIODOR

Other sources of heliodor include Minas Gerais, southeast Brazil; Rangkul in the Tian Shan Mountains, northeast Tajikistan; Litchfield County, Connecticut; Mursinsk, Urals, central Russia; and Nasarawa state, northern Nigeria.

▲ *Baguette-cut heliodor gem (top), cabochon-cut cat's-eye heliodor (below).*

◄ *Heliodor crystal.*

HELIODOR	
CHEMICAL COMPOSITION	$Be_3Al_2Si_6O_{18}$
COLOR	YELLOW, FROM ORANGE-YELLOW TO GREEN YELLOW
REFRACTIVE INDEX	1.57–1.6
RELATIVE DENSITY	2.80
HARDNESS	7.5–8
CRYSTAL GROUP	HEXAGONAL
CLEAVAGE	ONE DIRECTION, POOR
FRACTURE	CONCHOIDAL
TENACITY	BRITTLE
LUSTER	VITREOUS
TRANSPARENCY	TRANSPARENT TO TRANSLUCENT
DISPERSION	LOW, .014
BIREFRINGENCE	LOW, 0.004–.008
PLEOCHROISM	WEAK
LOCATION Brazil, Namibia, Nigeria, Russia, Tajikistan, United States.	

GOSHENITE

Brilliant

Step

Mixed

While most members of the beryl group, such as EMERALD or AQUAMARINE, are famous for their colors, goshenite is the clear variety of the gem. It is named for Goshen, Hamphire County, western Massachusetts, where it was originally found, but is also known as **white beryl** or **lucid beryl**, names that are more descriptive of its appearance.

Because the other varieties of beryl derive their color from impurities, it is tempting to believe that goshenite is colorless because it is pure, but this is not necessarily the case: some impurities present in the stone may actually inhibit the color that would normally be produced by other impurities.

Goshenite has been used since ancient times. The Greeks used it for lenses, and goshenite lenses made up the first spectacles. Because it facets and polishes well, it has also been long used as a gemstone. It is also a popular material for crystal balls, and some people believe it is a better conduit than QUARTZ.

On its own, goshenite is an attractive stone. The best specimens are free of inclusions, except those that create asterism. Large crystals are often available, which appeal to collectors.

Inexpensive because it is relatively abundant, goshenite has been used to imitate other gems. In its pure state, it is occasionally used as a substitute for DIAMOND. It is also used in composite stones, where colored cement or a thin slice of a colored stone is sandwiched between goshenite and then bezel mounted. Equally, it is found in foilback stones, in which colored foil, especially green (to imitate emerald) or silver (to imitate diamond), is placed behind goshenite, which is then mounted in a closed setting. Goshenite, like other beryls, can also be irradiated to produce a yellow or blue stone.

GOSHENITE

CHEMICAL COMPOSITION	$Be_3Al_2Si_6O_{18}$
COLOR	CLEAR
REFRACTIVE INDEX	**1.57–1.6**
RELATIVE DENSITY	**2.6–2.8**
HARDNESS	**7.5–8**
CRYSTAL GROUP	HEXAGONAL
CLEAVAGE	ONE DIRECTION, POOR
FRACTURE	CONCHOIDAL
TENACITY	BRITTLE
LUSTER	VITREOUS
TRANSPARENCY	TRANSPARENT
DISPERSION	LOW, .014
BIREFRINGENCE	LOW, .004–.008
PLEOCHROISM	NONE

LOCATION Brazil, Canada, China, Pakistan, United States.

GOSHENITE

Goshenite is found in most beryl locations. Beryl most commonly occurs as an accessory mineral in granites, and is usually found in cavities and in granite pegmatites. Beryl crystals in some pegmatites grow to very large sizes, even up to 30 feet (10 meters) in length. It also occurs in mica schists and gneisses in association with PHENAKITE, RUTILE, and CHRYSOBERYL.

▲ *Step-cut goshenite gem.*

▶ *Goshenite crystal.*

MORGANITE

Brilliant

Step

▶ *Scissors-cut morganite gem.*

Named after the banker and gemstone *aficionado* J. P. Morgan (1837–1913), morganite is the rare pink member of the beryl group of gems. Its color, which ranges from a yellowish salmon-pink, through champagne, to an almost lilac pink, is the source of its attraction to jewelry lovers and gem collectors.

Like other beryls, morganite commonly occurs as an accessory mineral in granites, and is usually found in cavities and in granite pegmatites. Because of its hardness, it is also sometimes found in alluvial deposits. It is associated with QUARTZ, albite, and muscovite mica, as well as other pegmatite accessory minerals. It is occasionally known as **vorobevite**.

Beryl is essentially colorless, but morganite derives its pink color from impurities of magnesium present in the stone. Iron impurities can give the stone a yellowish tinge. However, heating reduces the iron content, resulting in a purer pink stone.

Morganite is one of the rarest of the beryls, making it an expensive stone. Specimens should have a nice luster and not have any obvious inclusions. Faceting is important to bring out the luster. Unfaceted, morganite resembles ROSE QUARTZ, but cutting enhances its appearance. Morganite is also dichroic, showing pink from one angle and clear from another, and cutting must take that into consideration. Rarely, bicolored stones are found that are part-morganite and part-AQUAMARINE, but these are museum pieces.

MORGANITE

Morganite was discovered in Madagascar in 1991, and named by the gemmologist George Frederick Kunz (1856–1932). The Madagascar lode set the standard for the coloration of the gem, and its bright, lilac-pink color has not been equaled since. Other sources include Nuristan, Afghanistan; Minas Gerais, southeast Brazil; and Pala District, San Diego County, southern California.

▶ *Morganite crystal.*

MORGANITE	
CHEMICAL COMPOSITION	$Be_3Al_2Si_6O_{18}$
COLOR	PINK, FROM SALMON-PINK
	TO ALMOST LILAC
REFRACTIVE INDEX	1.57–1.6
RELATIVE DENSITY	2.71–2.9
HARDNESS	7.5–8
CRYSTAL GROUP	HEXAGONAL
CLEAVAGE	ONE DIRECTION, POOR
FRACTURE	CONCHOIDAL
TENACITY	BRITTLE
LUSTER	VITREOUS
TRANSPARENCY	TRANSPARENT
DISPERSION	LOW, .014
BIREFRINGENCE	LOW, .004–.008
PLEOCHROISM	NONE

LOCATION Afghanistan, Brazil, Italy, Madagascar, Mexico, Namibia, Pakistan, United States, Zimbabwe.

The rarest of the gems in the beryl group, red beryl is only found in three places, all in the United States: the Wah Wah Mountains and the Thomas Mountains of central Utah, and the Black Mountains of southwest New Mexico.

It occurs in the silica-rich volcanic rock known as topaz-bearing rhyolites. There, it is associated with QUARTZ, TOPAZ, SPESSARTINE, and HEMATITE, among others. It derives its red color from manganese, which is thought to substitute for aluminum in the beryl structure.

In 1904 the collector Maynard Bixby discovered the bright red stone in the Thomas Range in Juab County, Utah. The stone was named **bixbite** for him. However, this name is easily confused with bixbyite, a manganese iron oxide that is also mined in Utah but is not a gemstone. Because of this confusion, the gem is frequently known as red beryl, or even sometimes **red emerald**, linking it to its illustrious green cousin. But red beryl is more rare and more costly than emerald.

The best red beryl has a deep raspberry color and is free of obvious flaws or inclusions. Large gemstones are incredibly rare, and the average faceted stone is 0.15 carats, with faceted stones of more than 2 carats infrequently seen. The largest crystal yet recovered weighed about 54 carats, and the largest faceted gemstone to date weighed 8.0 carats. Because flaws and inclusions are common, the stone is challenging to cut, and clarity dramatically increases a specimen's value.

Brilliant

◀ *Brilliant-cut red beryl gem.*

▼ *Red beryl crystal on matrix.*

RED BERYL

The only gem-quality red beryl is mined in the Wah Wah Mountains, and productivity is very low; it is estimated that only one gemstone is recovered for every tonne of ore. Mining has taken place only since 1978. In fact, most crystals are not of gem quality, because they are too small to be faceted — most are under 2 inches (5 centimeters) in length.

RED BERYL

CHEMICAL COMPOSITION	$Be_3Al_2Si_6O_{18}$
COLOR	RED
REFRACTIVE INDEX	1.57–1.6
RELATIVE DENSITY	2.66–2.70
HARDNESS	7.5–8
CRYSTAL GROUP	HEXAGONAL
CLEAVAGE	INDISTINCT
FRACTURE	CONCHOIDAL
TENACITY	BRITTLE
LUSTER	VITREOUS
TRANSPARENCY	TRANSLUCENT TO OPAQUE
DISPERSION	MEDIUM, .014
BIREFRINGENCE	WEAK, .006–.009
PLEOCHROISM	DISTINCT, PURPLE-RED TO ORANGE-RED

LOCATION United States.

IOLITE

Step

Cabochon

Mixed

▶ *Faceted iolite gemstones.*

Renowned for its pleochroism, the mineral **cordierite** has both aesthetic and industrial applications. Although it was only officially named in 1812, for P. L. A. Cordier (1777–1861), the French geologist who described it, it has been used and admired for centuries.

In addition to being set in jewelry, cordierite also has industrial and scientific uses. It is said that the Vikings used a thin sliver of cordierite as a polarizing lens, a navigational tool that determines the exact position of the sun. Today, cordierite is used as an electrical insulator and in heating implements.

▲ *Slightly included cushion-cut iolite gem, 3.67 carats.*

Iolite is the variety of cordierite that is used as a gemstone. Its name derives from the Greek *io*, meaning "violet flower." Because it is common, it is an inexpensive gem, but its strong pleochroism makes it distinctive. Depending on the angle from which it is viewed, a stone can vary from bluish-purple to a wan gray-yellow. Sometimes it is also known as "water sapphire" because a clear stone can resemble a slightly violet SAPPHIRE. Iolites were very popular in jewelry in Europe in the 18th century, but today they are used infrequently. Durable if slightly brittle, they are usually cut into stepped rectangles. When cutting, it is important to orient the stone for color in order to enhance its color display. Beads of iolite are also fashioned.

IOLITE

Cordierite is a relatively common mineral found in aluminous rocks that have undergone medium- to high-grade contact or regional metamorphism. It is found in hornfelses, schists and gneisses in association with andalusite, spinel, quartz, and biotite mica. It is also found in igneous rocks that have assimilated aluminous sediments. The best gemstones come from Sri Lanka, India, Madagascar, and Burma (Myanmar).

▶ *Cordierite crystal on matrix.*

IOLITE	
CHEMICAL COMPOSITION	$(Mg,Fe)_2Al_4Si_5O_{18}$
COLOR	DARK BLUE, GRAYISH BLUE
REFRACTIVE INDEX	1.5
RELATIVE DENSITY	2.53–2.65
HARDNESS	7–7.5
CRYSTAL GROUP	ORTHORHOMBIC
CLEAVAGE	ONE: POOR CLEAVAGE, BASAL PARTING
FRACTURE	SUBCONCHOIDAL TO UNEVEN
TENACITY	BRITTLE
LUSTER	VITREOUS
TRANSPARENCY	TRANSPARENT TO TRANSLUCENT
DISPERSION	.017, WEAK
BIREFRINGENCE	.01
PLEOCHROISM	STRONG
LOCATION Bolivia, Brazil, Burma (Myanmar), Canada, Finland, Germany, Madagascar, Sri Lanka, Sweden, United States.	

TOURMALINE

Tourmaline is a complex ferromagnesian silicate that contains boron. It displays a greater range of color than any other gemstone, with different colored members OF the group having separate names. These include SCHORL (black), RUBELLITE (pink, red), INDICOLITE (blue), ACHROITE (colorless), DRAVITE (brown), buergerite (brown), chromdravite (green), siberite (violet), and uvite (black, brown, yellow-green). Often a crystal will have different colors at its two ends, or one color at the center and another outside.

Tourmaline was first discovered on the island of Elba, off the west coast of Italy. The name derives from the Sinhalese term *turmali* given to colored mixed crystals on the island of Sri Lanka. These were imported to Europe in the early 19th century as ZIRCONS, but were found to be a previously undescribed mineral. The true mineral name for most gem tourmaline is **elbaite**, named for its type locality.

Tourmaline commonly occurs in granite pegmatites, or in granites which have undergone metasomatism by boron-bearing fluids. It also occurs in sediments adjacent to such granites, and as an accessory mineral in schists and gneisses. The crystals are usually long thin prisms vertically striated. When heated or rubbed, a crystal acquires an electric charge and attracts small objects such as dust and hair. Tourmaline is used in electrical devices and to produce pressure gauges.

Colored crystals are used as gemstones when perfectly transparent. The most popular cuts are the mixed, trap, table, and step. Inclusions, such as black patches, are sometimes cut *en cabochon* to show a cat's-eye. Tourmaline is fashioned into jewelry like pendants, bracelets, rings, and earrings. It is also popular for beads and carved figurines.

▲ *Watermelon-tourmaline plaque enhanced with an old mine-cut diamond.*

▲ *Tourmalines can be made up of two or more colors: a pink and white gem (middle), and a pink and green watermelon tourmaline (bottom).*

◄ *Tourmaline gems come in a wide range of colors.*

163

Pendeloque

Step

Cabochon

The red or pink variety of TOURMALINE, rubellite is the rarest member of its gem family. In fact it is more rare than RUBY or red SPINEL, which it rivals in color. These three gems are the only ones that occur in a true deep red.

Rubellite, named for the Latin word for red, is a member of the **elbaite** group of tourmalines. Like other minerals in the group, it is a high-pressure, high-temperature mineral. It usually occurs in granite pegmatites, or in granites that have undergone metasomatism by boron-bearing fluids. It is also found in sediments adjacent to such granites, and as an accessory mineral in schists and gneisses. Rubellite is rich in lithium and free of magnesium and iron.

Rubellite can form crystals of up to 3 feet (1 meter) long. The crystals, often elongated, are trigonal and prismatic, with vertical striations. The red color of rubellite occurs because of impurities in the center of the stone. These impurities usually flaw the stone internally or cause it to crack. That is why good-quality rubellite is so rare.

Rubellite tends to be a bit pinker than ruby or red spinel. However, the similarity in appearance is strong enough that some of the red stones in the Russian crown jewels, for centuries believed to be rubies, are today thought to be rubellite. Some rubellite specimens may show a cat's-eye effect when cut as a cabochon, and the gem is strongly dichroic, which differentiates it from red spinel. A hard and durable stone, rubellite is suitable for jewelry and ornamental carving.

▲ *Rubellite crystals.*

RUBELLITE

Major sources of rubellite include Kunar, Nuristan, northeast Afghanistan; Minas Gerais, southeast Brazil; the Sahatany Valley, Antsirabe, central Madagascar; and Auburn, Maine, USA.

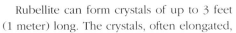

◄ *Brilliant-cut rubellite gem.*

RUBELLITE	
CHEMICAL COMPOSITION	$Na(Li,Al)_3Al_6(BO_3)_3$ $Si_6O_{18}(OH)_4$
COLOR	PINK TO RED
REFRACTIVE INDEX	1.62–1.64
RELATIVE DENSITY	3.03–3.10
HARDNESS	7–7.5
CRYSTAL GROUP	TRIGONAL
CLEAVAGE	NONE
FRACTURE	CONCHOIDAL
TENACITY	BRITTLE
LUSTER	VITREOUS
TRANSPARENCY	OPAQUE TO TRANSPARENT
DISPERSION	.017, MEDIUM
BIREFRINGENCE	.014–.024, MEDIUM
PLEOCHROISM	STRONG
LOCATION Afghanistan, Brazil, Germany, Madagascar, United States.	

Indicolite or indigolite is a deep, almost neon, blue gemstone variety of tourmaline. Its name derives from its distinctive color. It is a richer blue than many other blue stones, such as AQUAMARINE.

Indicolite is a gem variety of the **elbaite** mineral group. Like other minerals in the group, it is a high-pressure, high-temperature mineral. It commonly occurs in granite pegmatites, or in granites that have undergone metasomatism by boron-bearing fluids; it also occurs in sediments adjacent to such granites, and as an accessory mineral in schists and gneisses.

Indicolite is strongly pleochroic, and the stone appears darker as you look down the crystal. This must be considered during cutting. It is sometimes heat-treated to lighten the color. When cut as a cabochon, some indicolites show a cat's-eye effect. Its tranquil, blue color is said to heighten relaxation.

Step

▲ *Step-cut indicolite gem.*

Cabochon

◀ *Indicolite crystal.*

INDICOLITE

INDICOLITE	
CHEMICAL COMPOSITION	$Na(Li,Al)_3Al_6(BO_3)_3$ $Si_6O_{18}(OH)_4$
COLOR	BLUE
REFRACTIVE INDEX	1.62–1.68
RELATIVE DENSITY	3.03–3.10
HARDNESS	7–7.5
CRYSTAL GROUP	TRIGONAL
CLEAVAGE	NONE
FRACTURE	CONCHOIDAL
TENACITY	BRITTLE
LUSTER	VITREOUS
TRANSPARENCY	OPAQUE TO TRANSPARENT
DISPERSION	MEDIUM, .017
BIREFRINGENCE	MEDIUM, .014–.022
PLEOCHROISM	STRONG

LOCATION Brazil, Mexico, Namibia, Pakistan, Russia, United States.

INDICOLITE

Indicolite is a very rare form of tourmaline. Only small stones are available, and anything more than 1 carat is very expensive. A violet blue type of indicolite, known as **siberite**, is found in eastern Siberia, Russia. In 1989 a very precious neon-blue variety containing copper and a high gold content was discovered at Sao Jose de Batalha, Paraiba, Brazil.

Other sources of indicolite include Linópolis, Minas Gerais, southeast Brazil; the Erongo Mountains, Damaraland, northwest Namibia; and Gilgit, northern Pakistan.

◀ *Various cuts and colors of indicolite.*

ACHROITE

Brilliant

Mixed

Achroite is the colorless variety of tourmaline. The gem is extremely rare, and there is little demand for it. While some specimens are naturally occurring, it is also possible to generate achroite by heating pale pink, gemstone-quality tourmaline.

Achroite derives from the Greek word *achroos*, meaning "without color." It is a member of the **elbaite** group of tourmalines. Like other minerals in the group, it is a high-pressure, high-temperature mineral.

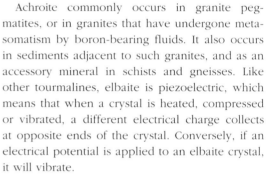

▲ *Oval cushion-cut achroite gem.*

Achroite commonly occurs in granite pegmatites, or in granites that have undergone metasomatism by boron-bearing fluids. It also occurs in sediments adjacent to such granites, and as an accessory mineral in schists and gneisses. Like other tourmalines, elbaite is piezoelectric, which means that when a crystal is heated, compressed or vibrated, a different electrical charge collects at opposite ends of the crystal. Conversely, if an electrical potential is applied to an elbaite crystal, it will vibrate.

Achroite is rarely used in jewelry. It is sometimes possible to find a stone that contains a spot of color surrounded by clear material, and these are particularly striking when cut appropriately. In general, because it is clear, achroite can be cut in any direction. In gemlore, its clarity is said to facilitate communication with the angelic realm.

Collectors should be aware that the term "achroite" can also be applied to other gems, such as achroite BERYL.

▲ *Achroite crystal in matrix.*

ACHROITE	
CHEMICAL COMPOSITION	Na(Li,Al)$_3$Al$_6$(BO$_3$)$_3$ Si$_6$O$_{18}$(OH)$_4$
COLOR	COLORLESS OR NEARLY SO
REFRACTIVE INDEX	1.62–1.64
RELATIVE DENSITY	3.03–3.10
HARDNESS	7–7.5
CRYSTAL GROUP	TRIGONAL
CLEAVAGE	NONE
FRACTURE	CONCHOIDAL
TENACITY	BRITTLE
LUSTER	VITREOUS
TRANSPARENCY	OPAQUE TO TRANSPARENT
DISPERSION	MEDIUM, .017
BIREFRINGENCE	MEDIUM, .014–.024
PLEOCHROISM	STRONG

LOCATION Afghanistan, Madagascar, United Kingdom, United States.

ACHROITE

The major source of achroite is the mines around St Austell, Cornwall, southwest England. It is also found in Androscoggin County and Oxford County, Maine, United States. Other localities include Afghanistan and Madagascar.

DRAVITE

Dravite is the least known of the gems in the tourmaline family, but deserves a higher profile. It makes a beautiful faceted gemstone with a hardness suited to jewelry, and it also forms large, well-shaped crystals that are an asset to any respectable gem collection.

Dravite is an end member of the tourmaline family, and it derives its color from its high magnesium content. Its name comes from the district of Drave, Styria, southeast Austria, where it was first identified. Like other varieties of tourmaline, it is a high-pressure, high-temperature mineral. It usually occurs in granite pegmatites, or in granites that have undergone metasomatism by boron-bearing fluids. It is also found in sediments adjacent to such granites, and as an accessory mineral in schists and gneisses.

While brown dravite is not uncommon, gem-quality stones are reasonably rare. Many tend toward the darker color variations, although heat treatment can lighten them. Like other tourmalines, dravite is strongly dichroic and must be cut to take advantage of the lighter, more attractive coloration. Because of its natural, earthlike coloring, dravite is said to help restore emotional balance and stability. Dravite crystals can be quite large.

Brilliant

Cushion

◄ *Brilliant-cut dravite gem.*

DRAVITE

Some of the best dravite crystals come from Yinnietharra in Western Australia. Other sources include Cochabamba, central Bolivia; Araçuaí, Minas Gerais, southeast Brazil; Ossola Valley, Piedmont, northwest Italy; eastern Siberia, Russia; Nora, Västmanland, eastern Sweden; and the Pamir Mountains, Tajikistan. Sites in North America include Sandon, British Columbia, Canada, and Oxford County, Maine, USA.

DRAVITE	
CHEMICAL COMPOSITION	$Na,Mg_3Al_6(BO_3)_3$ $Si_6O_{18}(OH)_4$
COLOR	LIGHT TO DARK BROWN
REFRACTIVE INDEX	1.62–1.68
RELATIVE DENSITY	3.03–3.15
HARDNESS	7–7.5
CRYSTAL GROUP	TRIGONAL
CLEAVAGE	NONE
FRACTURE	UNEVEN TO CONCHOIDAL
TENACITY	BRITTLE
LUSTER	VITREOUS
TRANSPARENCY	OPAQUE TO TRANSPARENT
DISPERSION	.017, MEDIUM
BIREFRINGENCE	.014–.024, MEDIUM
PLEOCHROISM	STRONG
LOCATION Australia, Bolivia, Brazil, Canada, Mexico, Tajikistan, United States.	

► *Rough pebble of dravite.*

Brilliant

Mixed

Although it is rarely used in jewelry, schorl, the black form of tourmaline, forms very attractive, long crystals, with a great number of crystal faces, which appeal to collectors.

Schorl is one of the end members of the tourmaline family. It is rich in sodium and derives its black color from its high iron content. Like other minerals in the group, it is a high-pressure, high-temperature mineral. It usually occurs in granite pegmatites, or in granites that have undergone metasomatism by boron-bearing fluids. It is also found in sediments adjacent to such granites, and as an accessory mineral in schists and gneisses.

In its pure form, schorl is not often cut because it is black, although it facets well like the other tourmaline gems. In the late 19th century, it was commonly used in mourning jewelry. Today, schorl is sometimes used in inlay work, since it is hard-wearing. When long, thin crystals of schorl are found as inclusions in clear, the resulting specimen is known as **tourmaline quartz**. With the black, needle-like crystals criss-crossing the clear stone, it is very attractive and often used for jewelry, ornaments and tumbled stones.

SCHORL

Schorl is widely distributed and is the most abundant member of the tourmaline group. Major deposits are found at South Australia and Victoria, Australia; Minas Gerais, southeast Brazil; Québec, Canada; Haut-Rhin, Alsace, northeast France; Fichtelgebirge, Bavaria, southern Germany; Sahatany Valley, Antsirabe, central Madagascar; Baltistan, northern Pakistan; Devon and Cornwall, southwest England. Sites in the United States include Androscoggin County and Oxford County, Maine; Cheshire County, New Hampshire; Middlesex County, Connecticut; and San Diego County, California.

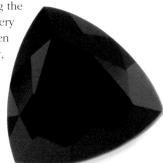

▶ *Mixed-cut schorl gem.*

◀ *Schorl crystals on matrix.*

SCHORL	
CHEMICAL COMPOSITION	$Na,Fe^{2+}{}_3Al_6(BO_3)_3$ $Si_6O_{18}(OH)_4$
COLOR	BLACK
REFRACTIVE INDEX	1.62–1.65
RELATIVE DENSITY	3.10–3.25
HARDNESS	7–7.5
CRYSTAL GROUP	TRIGONAL
CLEAVAGE	NONE
FRACTURE	UNEVEN TO CONCHOIDAL
TENACITY	BRITTLE
LUSTER	VITREOUS TO SUBMETALLIC
TRANSPARENCY	OPAQUE
DISPERSION	.017, MEDIUM
BIREFRINGENCE	.014–.024, MEDIUM
PLEOCHROISM	STRONG

LOCATION Australia, Brazil, Canada, Germany, Madagascar, Pakistan, United Kingdom, United States.

First identified on the slopes of Mount Vesuvius, southern Italy, vesuvianite has since been discovered to be widespread. While it is usually colorless, its colored varieties attract the attention of collectors and jewelry lovers alike.

Brilliant

Vesuvianite occurs in impure limestones that have undergone contact metamorphism, such as the blocks of dolomitic limestone erupted from Vesuvius. It is frequently accompanied by GROSSULAR GARNET, wollastonite, DIOPSIDE, and CALCITE. Massive varieties may easily be mistaken for GARNET, EPIDOTE, or diopside.

▲ *Baguette-cut vesuvianite gem.*

Baguette

The gemstone is also known as **idocrase**, from the Greek words *idea* (meaning "likeness") and *krasis* (meaning "mixture"), because its crystals show a mixture of other mineral forms. Visually it is often confused with garnet, AXINITE, PERIDOT, and varieties of TOURMALINE.

Mixed

One of the most popular varieties of vesuvianite is **californite**, a lumpy, green variety that resembles JADEITE. Usually cut as a cabochon, or carved into figurines or small bowls, it is also sometimes known as vesuvian jade. Another popular variety is blue **cyprine**, most of which is mined in Norway.

◄ *Vesuvianite crystals on matrix.*

▼ *Polished vesuvianite gem.*

The appearance of vesuvianite is enhanced by cutting. The more transparent stones are suitable for jewelry and are usually fashioned into table cuts or step cuts. Intact crystals of display size are rare and therefore very expensive. The crystals, usually short and thick, are striated along their length.

VESUVIANITE

CHEMICAL COMPOSITION	$Ca_{19}Fe(Mg,Al)_8Al_4$ $(SiO_4)_{10}(Si_2O_7)_4(OH)_{19}$
COLOR	USUALLY DARK GREEN OR BROWN, ALSO YELLOW
REFRACTIVE INDEX	1.70–1.75
RELATIVE DENSITY	3
HARDNESS	6–7
CRYSTAL GROUP	TETRAGONAL
CLEAVAGE	POOR
FRACTURE	SUBCONCHOIDAL TO UNEVEN
TENACITY	BRITTLE
LUSTER	VITREOUS TO RESINOUS
TRANSPARENCY	SUBTRANSPARENT TO TRANSLUCENT
DISPERSION	WEAK, .019
BIREFRINGENCE	.005
PLEOCHROISM	WEAK, DISTINCT IN THICK SECTIONS

LOCATION Italy, Norway, Romania, Switzerland, United States.

VESUVIANITE

Recently, a wide range of colors have been found including green specimens from California and Pakistan, yellow from New York, blue from Norway, and colorless from Siberia. Clear, purple vesuvianite crystals have been found in addition to the classic pink and green vesuvianite crystal clusters from the Jeffrey Mine, Asbestos, Québec, Canada, the world's largest asbestos mine.

ENSTATITE

Step

Cabochon

ENSTATITE

Major sources of enstatite include Québec, Canada; the Eifel Mountains, Rhineland-Palatinate, western Germany; Snarum, Buskerud, southern Norway; Swat Valley, Northwest Frontier Province, Pakistan; Kola Peninsula, northwest Russia; Embilipitiya, southwest Sri Lanka; Mogok, Mandalay, central Burma (Myanmar); Mbeya, southwest Tanzania; and the Lizard Peninsula, Cornwall, southwest England. Important US locations include Brewster, New York; Boulder, Colorado; Webster, Jackson County and Corundum Hill, Macon County, North Carolina; Lancaster County, Pennsylvania; Bare Hills, Maryland; and Whatcom County, Washington.

▲ *Mixed-cut enstatite gem.*

Enstatite lies at the magnesium-rich end of the orthopyroxene series. These minerals are those members of the pyroxenes that crystallize in the orthorhombic system. At the other end of the orthopyroxene series lies the very rare, iron-rich mineral ferrosilite, $(Fe^{2+},Mg)_2Si_2O_6$. The name hypersthene is often applied to intermediate members of this series. **Bronzite** is a bronze-colored variety of enstatite with a hypersthene component.

Enstatite is a common constituent of igneous rocks such as gabbro and pyroxenite. It also occurs in some andesitic volcanic rocks and in stony meteorites. Enstatite is often found in association with olivine (gem variety PERIDOT), phlogopite, clinopyroxene, DIOPSIDE, SPINEL, and PYROPE. Its name derives from the Greek *enstates*, meaning "opponent" – a reference to its refractory nature.

Enstatite is often pale green, but can also be light brown, colorless, white or gray. It is generally found in the form of masses, or in fibrous or lamellar aggregates, but is occasionally found as short prismatic crystals. When traces of chromium are present, the emerald-green variety **chrome-enstatite** is formed, which is faceted as a gemstone. However, most enstatite is polished and cut as a cabochon.

ENSTATITE	
CHEMICAL COMPOSITION	Mg_2Si_2O6
COLOR	PALE GREEN TO DARK BROWNISH GREEN
REFRACTIVE INDEX	1.65–1.68
RELATIVE DENSITY	3.1–3.3
HARDNESS	5–6
CRYSTAL GROUP	ORTHORHOMBIC
CLEAVAGE	TWO DISTINCT
FRACTURE	UNEVEN
TENACITY	BRITTLE
LUSTER	VITREOUS
TRANSPARENCY	TRANSLUCENT, RARELY TRANSPARENT
DISPERSION	WEAK TO MODERATE
BIREFRINGENCE	.007–.020
PLEOCHROISM	DISTINCT

LOCATION Burma (Myanmar), India, Norway, Sri Lanka, Tanzania, United Kingdom, United States.

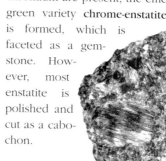

▶ *Enstatite crystals on matrix.*

Hypersthene is a common **pyroxene** rich in iron. The name derives from the Greek words *hyper*, meaning "above," and *stenos*, meaning "strength," because hypersthene is harder than hornblende, a mineral with which it is often confused. Hypersthene is in a series of minerals with ENSTATITE, and these two similar members are sometimes listed as enstatite-hypersthene. The deep color of hypersthene is due to the 50% content of iron (with inclusions of HEMATITE and goethite), whereas enstatite is magnesium rich. Other series members include **bronzite** and **ferrosilite**.

Hypersthene is found in igneous and metamorphic rocks, as well as meteorites. It is not industrially mined. Associated minerals include iron, biotite, QUARTZ, FELDSPARS, olivine (gemmy PERIDOT), and ALMANDINE. The crystals are usually massive, or found in coarse lamellar or fibrous aggregates. They are often too dark to facet but may be fashioned into cabochons to highlight their sparkling inclusions.

Baguette

Cushion

◀ *Brilliant-cut hypersthene gem.*

HYPERSTHENE

Gem-quality hypersthene is found in Germany, the United States, and a few other countries including Mexico, Russia and Sweden, but gemstones are not numerous or well known. Sources include the Bavarian Forest, southern Germany; Harz Mountains, Lower Saxony, northwest Germany; and the Eifel Mountains, Rhineland-Paltinate, western Germany. US locations include the Adirondack mountain region of New York, especially at North Creek; Crestmore, California; Comanche County, Oklahoma; and Los Alamos County, New Mexico. In the United Kingdom, deposits exist on Thurstaston Beach, Wirral, Merseyside, northwest England. Other deposits of hypersthene include Nuuk, Greenland; Enderby Land, Antarctica; Filipstad, Värmland, eastern Sweden; and Zimapán, Hidalgo, southern Mexico.

HYPERSTHENE	
CHEMICAL COMPOSITION	$(Mg, Fe)_2Si_2O_6$
COLOR	PALE GREEN TO GRAYISH BLACK AND BROWN
REFRACTIVE INDEX	1.65–1.67
RELATIVE DENSITY	3.35
HARDNESS	5–6
CRYSTAL GROUP	ORTHORHOMBIC
CLEAVAGE	GOOD
FRACTURE	UNEVEN
TENACITY	——
LUSTER	VITREOUS
TRANSPARENCY	TRANSLUCENT TO OPAQUE
DISPERSION	WEAK
BIREFRINGENCE	.01–.02
PLEOCHROISM	DISTINCT

LOCATION Antarctica, Canada, Germany, Greenland, Mexico, Sweden, United Kingdom, United States.

▶ *Bronzite crystals on matrix.*

DIOPSIDE

Brilliant

Baguette

Step

Cabochon

Diopside, calcium magnesium silicate, is widely distributed around the world, occurring in contact-metamorphosed impure limestones and skarns and, more rarely, in basaltic igneous rocks. It is even found in meteorites.

Diopside falls at one end of the clinopyroxene mineral series, which ranges from diopside, $CaMgSi_2O_6$, to hedenbergite, $CaFeSi_2O_6$. Members of the series differ in the relative proportions of magnesium and iron ions they contain (though other ions make an appearance in other members of the series), with diopside being at the high-magnesium end. All members of the series belong to the monoclinic system. The clinopyroxenes are in turn members of the pyroxene group.

Crystals of diopside are short and columnar, with a square or eight-sided cross-section, often twinned and often forming striking-looking assemblages. Some specimens are fluorescent.

An important chromium-rich variety of diopside is called **chrome diopside**, and is noted for its deep green color. A rare blue variety called **violan** is found in some parts of Italy. There is also a green cat's-eye variety, owing the effect to inclusions of rutile needles. **Star diopside** also has rutile inclusions, producing a star with four rays.

Diopside is generally whitish to yellowish, or light green or light blue, and is transparent or translucent. The crystals have a perfect cleavage in two directions. The fracture is uneven. The mineralogist can recognize diopside in the field by the form of its crystals, its color, its fracture and cleavage, and its white or white-green streak.

▶ *Brilliant-cut diopside from South Africa.*

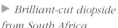

DIOPSIDE

Fluorescent diopside is found at Hastings County and Renfrew County, Ontario, Canada, while chrome diopside is mostly mined in Yakutia, eastern Siberia. Other sources of diopside include the Erzgebirge, Saxony, eastern Germany; the Valle d'Aosta, northwest Italy, and Mount Vesuvius area, southern Italy; Kola Peninsula, northwest Russia, and Lake Baikal, southern Siberia, Russia; Ludvika, Dalarna, central Sweden, and Nora, Västmanland, eastern Sweden. Major US sources include Sussex County, New Jersey.

▶ *Diopside crystals on matrix.*

DIOPSIDE	
CHEMICAL COMPOSITION	$CaMgSi_2O_6$
COLOR	GRAYISH WHITE TO LIGHT GREEN; TRANSLUCENT TO OPAQUE
REFRACTIVE INDEX	1.66–1.72
RELATIVE DENSITY	3.2–3.6
HARDNESS	5–6
CRYSTAL GROUP	MONOCLINIC
CLEAVAGE	PRISMATIC, GOOD
FRACTURE	UNEVEN
TENACITY	BRITTLE
LUSTER	VITREOUS
TRANSPARENCY	TRANSLUCENT TO OPAQUE
DISPERSION	WEAK TO DISTINCT
BIREFRINGENCE	.02–.03
PLEOCHROISM	DISTINCT
LOCATION Canada, Germany, Italy, Russia, Sweden, United States.	

Crystal and various cuts and colors of danburite.

Brilliant

Step

Mixed

Relatively new to the market and thus little known, danburite is a stone that is destined to become more popular. It has many good qualities as a gem, both for jewelry and for collections.

In jewelry, clear danburite makes a good alternative to DIAMOND, harmonizing well with other gems. However, with a hardness of 7–7.5, it becomes scratched and worn far sooner than a diamond. Collectors are attracted by its crystal clusters – the most sought after are the prismatic crystals from San Luis Potosí in north-central Mexico. Crystal enthusiasts believe that it is a good healing stone, particularly the pink-tinted specimens.

◀ *Triangular fancy-cut danburite gem.*

DANBURITE

Danburite is named for the town of Danbury, Fairfield County, southeast Connecticut, where it was first discovered in 1839. That source has since been buried beneath the suburban sprawl, but danburite has since been found around the world. It is associated with QUARTZ, FELDSPAR, CASSITERITE, DOLOMITE, and RUBY.

DANBURITE

CHEMICAL COMPOSITION	$Ca[B_2Si_2O_8]$
COLOR	USUALLY CLEAR, BUT SOMETIMES YELLOW OR PINK
REFRACTIVE INDEX	1.63–1.64
RELATIVE DENSITY	2.97–3.02
HARDNESS	7–7.5
CRYSTAL GROUP	ORTHORHOMBIC
CLEAVAGE	POOR IN ONE DIRECTION, BASAL
FRACTURE	UNEVEN TO CONCHOIDAL
TENACITY	BRITTLE
LUSTER	VITREOUS
TRANSPARENCY	TRANSPARENT TO TRANSLUCENT
DISPERSION	MEDIUM, .017
BIREFRINGENCE	WEAK, .006
PLEOCHROISM	WEAK
LOCATION	Burma (Myanmar), Japan, Mexico, Switzerland, United States.

▶ *Danburite crystals.*

JADEITE

▲ *Polished slice of jadeite.*

Jade has been cherished for millennia by cultures as far apart as China and Central America, and British Columbia (western Canada) and New Zealand.

The lustrous green stone, stronger than steel, has been used for tools and ornaments, and is still believed to confer luck, health and spiritual well-being upon its owners.

Today, we know that there are actually two separate gemstones known popularly as jade – jadeite and NEPHRITE – but this discovery was not made until 1863. The two gems are very similar: both are aluminum silicates; both are made of intertwined crystals that were compressed under high pressure; and they are often found side by side. Now that the distinction is known, the two can be distinguished visually. Broadly, nephrite is the more oily of the two, and jadeite occurs in a more vivid green as well as in a wider range of colors, including lavender, white, red, yellow, brown, orange, and pink.

Jadeite is formed at high pressures. As a result, the crystals are densely packed inside, accounting for its high relative density and its extreme strength. Because it is so strong, and also holds a cutting edge well, it was often used for weapons and tools. There are few *in situ* sources of jadeite – most is found as boulders, weathered to brown on the exterior, or in river gravel. When jadeite is traded, it is as huge boulders, perhaps with a small window cut in one area, and purchasers are gambling that the boulder, when split, will reveal good-quality jadeite.

Jadeite was widely used and revered by all the Central American cultures. The Olmecs, Mayans, Toltecs, and Aztecs created tools as well as beautifully carved ceremonial objects of jadeite. It was considered a magical stone, which could bring protection from the gods but also wealth and good fortune. The Aztecs even instituted a tax payable in jade.

It was in South and Central America that the Spanish *conquistadors* discovered jadeite and brought it back to Europe, where it was then virtually unknown.

Its name derives from the Spanish phrase *piedra de ijada*, meaning "stone of the side," referring to the belief that jade could cure kidney ailments and hip problems.

Jadeite was also brought to Europe by the Portuguese, who obtained it from their trading bases in China. Jade has equally ancient roots in China, where arguably it has been worked with the greatest skill in the world. Artifacts, both ornamental and utilitarian, have been

JADEITE	
CHEMICAL COMPOSITION	$Na(Al,Fe^{3+})Si_2O_6$
COLOR	USUALLY GREEN, SOMETIMES WHITE, LAVENDER OR RED
REFRACTIVE INDEX	1.66-1.68
RELATIVE DENSITY	3.25-3.36
HARDNESS	6.5-7
CRYSTAL GROUP	MONOCLINIC
CLEAVAGE	POOR IN TWO DIRECTIONS
FRACTURE	SPLINTERY TO UNEVEN
TENACITY	VERY TOUGH
LUSTER	VITREOUS
TRANSPARENCY	TRANSLUCENT TO OPAQUE
DISPERSION	NONE
BIREFRINGENCE	.02
PLEOCHROISM	NONE

LOCATION Burma (Myanmar), Canada, Guatemala, Japan, Kazakhstan, Mexico, Russia, United States.

◄ *Art Deco hairpin of carved and pierced jade.*

found dating back as early as the 18th century BC. For centuries, the Chinese preferred their native nephrite, mined in the Kunlun Mountains of Xinjiang, northwest China. When jadeite was first introduced from Burma (Myanmar) in the 16th century, Chinese artists and collectors rejected it, calling it disparagingly the "kingfisher stone," in reference to the many colors in which jadeite is available. It was not until the reign (1735–96) of the Qing Emperor Qianlong that jadeite was accepted in China – and this was the period during which jade work reached a new artistic peak.

▲ *Cabochon cut jadeite gem.*

Bead

Cameo

Polished

Today, jade remains one of the most popular gemstones. Jadeite, the rarer variety, is also the most prized, largely because of its better coloration. When it is found in a rich emerald green color, it is known as **imperial jade**, and is the most highly prized; interestingly, imperial jade derives its green color from the presence of chromium, the same element that gives EMERALD its vivid green. **Lavender jade**, its color caused by traces of iron, is rare and thus valuable; the rarest variety is when a green swirl appears in a white stone, a phenomenon known as "moss in snow."

Jadeite is selected based on its color, which should be free of brown or gray and should also be vibrant and vivid. The stone should be clear and free of irregularities. The more translucent the specimen is, with a honey-like appearance, the more expensive it is; virtually transparent stones are the most valuable. The best-quality samples are reserved for shaping into cabochons, while samples with some faults are reserved for carving, which can take advantage of any irregularities. Jade's strength makes it hard to work, since it cannot be chiseled. In fact, when diamonds were first introduced into China, they were prized for their ability to cut jade and not valued for their own beauty. Today, hard-wearing abrasives are easily available and are used to shape it.

Jadeite can be dyed, often green or lavender, but such dyed specimens are usually easily detected thanks to their unusually uniform appearance.

JADEITE

Jadeite occurs as grains in metamorphosed sodic sediments and volcanic rocks, and is associated with glaucophane and aragonite. It is intermediate in composition to albite and nepheline, although they have nothing in common in terms of appearance. Burma (Myanmar) is the major source of jadeite, and is the only source of **red jadeite**, although good stones are also still unearthed in Guatemala.

▲ *Mughal or Deccani carved jade bowl.*

Bead

Cameo

Polished

◀ *Nephrite finial carving of birds and flowers.*

K nown as the Yu stone, nephrite jade was the "Stone of Heaven" in ancient China. In archaic burial rites, six specific nephrite jade sculptures were buried with an individual to ensure his passage to Heaven. Nephrite jade was excavated from the Kunlun Mountains of northwest China, from 5000 BC to the 1700s, and China remains an important source for the mineral.

For centuries, nephrite and JADEITE were considered one and the same, and it was not until 1863 that they were identified as different minerals with similar appearance and properties.

Nephrite is a form of amphibole, a type of rock-forming silicate that is widely distributed in igneous and metamorphic rocks. It is usually an actinolitic or tremolitic amphibole. Amphiboles are typically fibrous, and this explains nephrite's durability: its fibrous crystals interlock, making it even stronger than steel. Its color varies from white through yellow to green, and ultimately to brown or black as more and more iron is substituted for magnesium in its structure. The most common impurities are the SPINEL minerals, which may undercut or cause pitting.

Nephrite is the toughest of all natural stones – it is so strong that it cannot be chiseled but has to be ground using sharp abrasives. In antiquity, it was often used in weaponry because of its great strength. Today, nephrite has a multiplicity of uses. Because it is less translucent than jadeite, it is better suited to delicately carved jewelry, such as cameos. Its durability suits it to all kinds of carving and also to architectural applications such as laminates. Less than 0.05% of nephrite extracted is of gem quality, with the rest suited to carving or building applications.

Nephrite is rarely treated. It is less likely to take up dye or stains than jadeite, although it may be waxed or oiled. Older pieces tend to lose some of their luster and may benefit from polishing.

NEPHRITE

China is not the only venerable source of nephrite. In Russia, it has been mined and crafted since 3000 BC. Tsar Alexander III's (1845–94) sarcophagus was carved from nephrite. For about 3,000 years, the stone has been highly prized by the Native North Americans of British Columbia, Canada, where it was known as *squa* or *lisht*. In New Zealand, where some of the best-quality nephrite is mined, it is sometimes known as "greenstone." For centuries, the Maori have made beautiful nephrite carvings, and they relied on it for tools until the Europeans introduced metals in the 18th century.

▶ *Nephrite polished pebble.*

NEPHRITE	
CHEMICAL COMPOSITION	$Ca_2(Mg,Fe)_5$ $[(OH,F)_2|Si_4O_{11}]_2$
COLOR	GREEN, WHITE, YELLOW, BROWN TO BLACK
REFRACTIVE INDEX	1.61–1.63
RELATIVE DENSITY	2.9–3.1
HARDNESS	6.5
CRYSTAL GROUP	MONOCLINIC
CLEAVAGE	NONE
FRACTURE	HACKLY
TENACITY	VERY TOUGH (ELASTIC)
LUSTER	VITREOUS
TRANSPARENCY	TRANSLUCENT TO OPAQUE
DISPERSION	NONE
BIREFRINGENCE	WEAK, .003
PLEOCHROISM	STRONG
LOCATION	Australia, Burma (Myanmar), Canada, China, Japan, New Zealand, Russia, United States.

Brilliant

Pendeloque

Some of the newest minerals to the market, the gemstone varieties of spodumene were only discovered around the turn of the 20th century: green hiddenite was identified in 1879 and lilac kunzite in 1902.

Spodumene occurs typically in lithium-bearing granite pegmatites and it is a major ore of lithium, the lightest of metals. It is found in association with other minerals including TOURMALINE and BERYL. Spodumene is mainly clear to grayish white, and its name comes from the Greek word *spodumenos*, meaning "burnt to ashes." Spodumene has been known to form very large crystals, some as long as 50 feet (15 meters) in length and weighing up to 90 tons. The prism faces are vertically striated lengthwise. Gem-quality spodumene is rarely found in large crystals.

▲ *Kunzite crystal.*

Step

Kunzite, named for the gemmologist George Frederick Kunz (1856–1932), is the better-known gemstone variety of spodumene. Its lilac color, due to the presence of manganese, is unique in the gem world.

Hiddenite is much less common, and its color, caused by the presence of chromium and iron, varies from yellowish-green to blue-green, and even sometimes a clear green that approaches the color of EMERALD. The best stones show a deep color.

Both kunzite and hiddenite are distinguished by their strong pleochroism. The top and bottom of a crystal tend to reveal the deepest colors. This color shift represents a challenge to cutters, as does the stone's brittleness and cleavage. Both gem varieties are usually cut into brilliants or table cuts. They are best suited to pins, earrings or pendants, because the stones are too fragile to adorn rings.

SPODUMENE	
CHEMICAL COMPOSITION	$LiAlSi_2O_6$
COLOR	USUALLY WHITE OR GRAYISH-WHITE, BUT ALSO GREEN OR LILAC
REFRACTIVE INDEX	1.66–1.67
RELATIVE DENSITY	3.17–3.23
HARDNESS	6–7
CRYSTAL GROUP	MONOCLINIC
CLEAVAGE	PRISMATIC, PERFECT
FRACTURE	UNEVEN, SPLINTERY
TENACITY	BRITTLE
LUSTER	VITREOUS
TRANSPARENCY	TRANSPARENT TO TRANSLUCENT
DISPERSION	WEAK, .017
BIREFRINGENCE	.015
PLEOCHROISM	STRONG

LOCATION Afghanistan, Brazil, Burma (Myanmar), Canada, Madagascar, Mexico, Namibia, Pakistan, Russia, Sweden, United States.

SPODUMENE

Spodumene was first discovered in California, which remains a good source of the gem, along with Brazil, Madagascar, and Pakistan. The main source of hiddenite is the Stony Point Mine in North Carolina, USA, where the gem was discovered by W. E. Hidden in 1879.

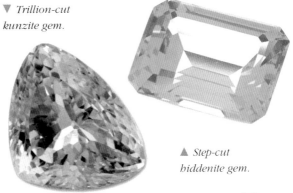

▼ *Trillion-cut kunzite gem.*

▲ *Step-cut hiddenite gem.*

RHODONITE

Bead

Cabochon

Cameo

Taking its name from the Greek word *rhodos*, meaning "pink," rhodonite is popular with jewelry enthusiasts and gem collectors alike. Its rose-pink body is laced with black veins of manganese oxide, called dendrites, which form bands, dots or intricate, web-like patterns. In color, rhodonite is similar to rhodochrosite, but rhodonite is distinguished by its dendrites. In addition, it is harder wearing and more acid resistant than rhodochrosite, and thus better for jewelry. Initially it may be black on the exterior, due to oxidation, but its characteristic pink tone appears once it is polished.

▲ *Rhodonite bead.*

Rhodonite was first discovered in the 17th century. During the 18th century it was extremely popular at the Russian court and was for many years considered the national stone of Russia, used not only in jewelry but also for architectural ornamentation, such as wall panel-ing. Transparent crystals are rare and highly sought after by collec-tors; they are extremely fragile and have to be handled delicately. They are sometimes fashioned into table cuts or brilliants. Most collec-tors value polished slices, which display good patterning. Samples can also be carved into figurines. For jewelry, rhodonite is cut as cabochons or as beads. It may also be used in cameos. Many different meanings have been attributed to the stone. It is said to be calming and soothing, and to imbue the wearer with con-fidence in matters of the heart. It is also sometimes called the "singer's stone" because it is said to enhance sensitivity to sound.

RHODONITE

Rhodonite commonly occurs in associ-ation with manganese ore deposits in hydrothermal or metasomatic veins, or else in regionally metamorphosed manganese-bearing sediments. Major deposits are found in Broken Hill, New South Wales, Australia, and the Urals of Russia, but it is widely distributed.

▶ *Rhodonite crystal on matrix.*

RHODONITE	
CHEMICAL COMPOSITION	$(Mn_2+,Fe_2+,Mg,Ca)SiO_3$
COLOR	PINK TO BROWN
REFRACTIVE INDEX	1.7
RELATIVE DENSITY	3.37–3.82
HARDNESS	5-6
CRYSTAL GROUP	PRISMATIC OR TABULAR
CLEAVAGE	IN THREE DIRECTIONS, ONE PERFECT, ONE GOOD
FRACTURE	CONCHOIDAL TO UNEVEN
TENACITY	BRITTLE, BUT LUMP VARIETIES ARE TOUGH
LUSTER	VITREOUS
TRANSPARENCY	TRANSPARENT TO TRANSLUCENT
DISPERSION	NONE
BIREFRINGENCE	.014
PLEOCHROISM	DISTINCT

LOCATION Australia, Canada, Costa Rica, Germany, Italy, Mexico, Peru, Russia, Sweden, United Kingdom, United States.

Pectolite's name derives from the Greek words *pektos* ("compacted"), because of its compacted structure, and *lithos* ("stone"). Fine specimens do occur, but the most desirable variety, a pale to sky-blue color, was not discovered until 1974 in the Bahamas and Dominican Republic. It is actually a rock mostly composed of pectolite. Sold as a gemstone under the tradename of **Larimar**, it often has a white spider-veined appearance. It looks like TURQUOISE, and is popular for jewelry items, including some modern Native American pieces.

Pectolite occurs typically, along with zeolites, in cavities in basalts and similar rocks. It forms aggregates of fibrous or acicular crystals, often radiating or stellate. The brittle splinters are sharp and must be handled carefully. Pectolite is often confused with other similar looking minerals, such as artinite, wollastonite, and okenite. Most of these, however, do not form with zeolites as pectolite tends to. Associated minerals include annite, apophyllite, amphibole, microcline, SODALITE, VESUVIANITE, poudretteite, CALCITE, PREHNITE, datolite, and SERPENTINE.

Cabochon

◄ *Pectolite crystal on matrix.*

PECTOLITE	
CHEMICAL COMPOSITION	NaCa$_2$Si$_3$O$_8$OH
COLOR	WHITE, GRAY, BLUE
REFRACTIVE INDEX	1.60–1.64
RELATIVE DENSITY	2.8–2.9
HARDNESS	4.5–5
CRYSTAL GROUP	TRICLINIC
CLEAVAGE	TWO PERFECT CLEAVAGES
FRACTURE	UNEVEN
TENACITY	BRITTLE
LUSTER	SILKY WHEN FIBROUS, OTHERWISE VITREOUS
TRANSPARENCY	SUBTRANSLUCENT TO OPAQUE
DISPERSION	PERCEPTIBLE
BIREFRINGENCE	.037
PLEOCHROISM	NONE
LOCATION Bahamas, Canada, Dominican Republic, Italy, South Africa, UK, United States.	

PECTOLITE

Pectolite's type locality is Mount Baldo in the Trentino-Alto Adige region of northeast Italy. Other locations include Labrador, Newfoundland, Canada; Kola Peninsula, northwest Russia; North Cape Province, South Africa; and Durham, northeast England. US sites include Pulaski County, Arkansas, and northern New Jersey. **Larimar** is only found in the Bahamas and Dominican Republic.

► *Cabochon-cut pectolite gem.*

Cushion

Step

Alithium aluminum silicate, petalite is not the most popular mineral but it is very collectable. Its name comes from the Greek words *petalon*, meaning "leaf," an allusion to its distinctive leaflike cleavage, and *lithos*, meaning "stone." Petalite crystals are rare and small, seldom reaching 3 carats. They are very fragile and seldom cut, though cushion mixed cuts are seen. Too brittle for daily wear in rings, the crystals are used for earrings, necklaces, and pendants. Massive petalite is cut as cabochons.

Petalite was discovered by Brazilian mineralogist J. B. de Andrada e Silva (1763–1838) in 1800. It occurs typically in lithium-bearing granite pegmatites alongside minerals such as SPODUMENE, TOURMALINE, LEPIDOLITE, and FELDSPARS. Lithium was discovered in 1817 by Swedish chemist Johann Arfvedson during an analysis of petalite ore. Petalite resembles cleavage masses of feldspar and is often distinguishable only by optical tests, although in flame tests it gives the red flame characteristic of lithium. Crystals have a glassy appearance and occur as tabular or columnar prisms.

PETALITE	
CHEMICAL COMPOSITION	$LiAlSi_4O_{10}$
COLOR	WHITE, GRAY, GREEN, COLORLESS, REDDISH
REFRACTIVE INDEX	1.50–1.51
RELATIVE DENSITY	2.4–2.5
HARDNESS	6–6.5
CRYSTAL GROUP	MONOCLINIC
CLEAVAGE	PERFECT
FRACTURE	SUBCONCHOIDAL
TENACITY	BRITTLE
LUSTER	VITREOUS; PEARLY ON CLEAVAGE SURFACE
TRANSPARENCY	TRANSPARENT TO TRANSLUCENT
DISPERSION	NONE
BIREFRINGENCE	.012
PLEOCHROISM	NONE

LOCATION Afghanistan, Australia, Brazil, Finland, Italy, Mozambique, Namibia, Russia, Sweden, United Kingdom, United States, Zimbabwe.

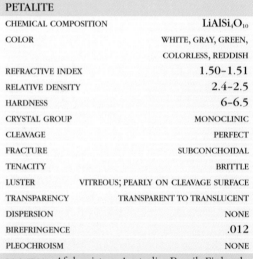

► *Petalite crystal fragment.*

▲ *Step-cut petalite gem from Brazil.*

PETALITE

Petalite's type locality is Utö, Haninge, Stockholm, Sweden. Other sources of petalite include Nuristan, northeast Afghanistan; Nepean, Western Australia; Araçuai, Minas Gerais, southeast Brazil; Alto Ligonha, Zambezia, central Mozambique; Karibib, eastern Namibia; and the Buryatia Republic, eastern Siberia. It is also found in Ontario and Manitoba, Canada. Sites in the US include Oxford County, Maine; Hampden County, Massachusetts; Middlesex County, Connecticut; Strafford County, New Hampshire; Cleveland County, North Carolina; Taos County, New Mexico; and Fremont County, Wyoming.

A rare member of the mica group, lepidolite has become available in large amounts to collectors in only the last decade. Jewelers are now cutting more stones, often in cabochons, and the pink to lilac shades of lepidolite gems are becoming more common. Color is the only way to visually identify the gem from other micas, and pale lepidolite is often confused with pink muscovite. A popular ornamental stone is made up of pink lepidolite and red TOURMALINE. Lepidolite is carved into small objects, such as ashtrays and vases. Mineral collectors also value single books (plates), especially of the violet variety.

This mica is a major source for lithium. Its name derives from the Greek words *lepidion,* meaning "scale," because specimens can look scaly, and *lithos,* meaning "stone." It is made up of hexagonal sheets of lithium aluminum silicate bonded weakly by layers of potassium ions, which give lepidolite its perfect cleavage. Such a rock is called "slaty" for its slatelike cleavage. The sheets are elastic and will return to their original shape after being bent. Lepidolite occurs in granite pegmatites often in association with TOURMALINE, SPODUMENE, QUARTZ, elbaite, TOPAZ, CASSITERITE, AMBLYGONITE, and FELDSPARS.

Cabochon

◄ *Lepidolite cabochon-cut gem.*

LEPIDOLITE

Lepidolite's type locality is Rožna, Bohemia, Czech Republic. Major sources of lepidolite include Kunar Province, Nuristan, northeast Afghanistan; Araçuaí, Minas Gerais, southeast Brazil; Bernic Lake, Manitoba, and Fredericton, New Brunswick, Canada; Vaasa, western Finland; Fichtelgebirge, Bavaria, southern Germany; Ensenada, Baja California, Mexico; and Karibib, eastern Namibia. US sites include San Diego County, California; Gunnison County, Colorado; Middlesex County, Connecticut; and Auburn and Oxford County, Maine.

LEPIDOLITE

CHEMICAL COMPOSITION	$K(Li,Al)_3(Si,Al)_4$ $O_{10}(OH,F)_2$
COLOR	PALE LILAC TO GRAY OR PALE PINK; ALSO COLORLESS
REFRACTIVE INDEX	1.55
RELATIVE DENSITY	2.8–2.9
HARDNESS	2.5–4
CRYSTAL GROUP	MONOCLINIC
CLEAVAGE	BASAL PERFECT
FRACTURE	UNEVEN
TENACITY	ELASTIC
LUSTER	VITREOUS; PEARLY ON CLEAVAGE SURFACES
TRANSPARENCY	TRANSPARENT TO TRANSLUCENT
DISPERSION	WEAK
BIREFRINGENCE	.029–.038
PLEOCHROISM	COLORLESS

LOCATION Afghanistan, Argentina, Brazil, Canada, Czech Republic, Finland, Germany, Madagascar, Mexico, Norway, Russia, Sweden, United States, Zimbabwe.

◄ *Lepidolite crystal.*

SERPENTINE

Cameo

Polished

Cabochon

Named for the snakeskin it is said to resemble, serpentine has been used for ornaments and jewelry for thousands of years, because it is suitable for carving and polishes to an attractive gleam. It may be confused with JADEITE or NEPHRITE, because of its similar coloration, and is sometimes known as "Korean jade" or "new jade." It is softer and scratches more easily than true jade.

The term serpentine encompasses a group of related minerals that are widely distributed around the world. It is a secondary mineral formed from minerals such as OLIVINE (gem PERIDOT) and ortho-pyroxene. It occurs in igneous rocks containing these minerals but typically in serpentinites, which have been formed by the alteration of olivine-bearing rocks. There are two basic forms of serpentine minerals: the **antigorites** and the **chrysotiles**. While antigorites are flaky, chrysotiles tend to be fibrous and are the source of asbestos.

Although serpentine is related to asbestos, which is a carcinogen, it does not represent a cancer risk unless it is fibrous. However, it is recommended that asbestos serpentines should be displayed in sealed containers.

Serpentine is used in many different ways. It may be carved into figurines or bowls, used in inlay work or as wall facings or even countertops. As jewelry, the stones may be carved into cabochons or into flat plates, which can be semitransparent. Because it is soft, it is not suitable for rings. Serpentine amulets have historically been worn to protect against snakebite or poisoning. The gem is also said to stimulate success and courage.

SERPENTINE	
CHEMICAL COMPOSITION	$Mg_6[(OH)_8|Si_4O_{10}]$
COLOR	GRAYISH-GREEN, YELLOW, BROWN
REFRACTIVE INDEX	1.56 (MEAN)
RELATIVE DENSITY	2.53–2.65
HARDNESS	2.5–4.0
CRYSTAL GROUP	MONOCLINIC
CLEAVAGE	MOOD
FRACTURE	CONCHOIDAL, UNEVEN
TENACITY	SOFT TO SPLINTERY
LUSTER	GREASY, WAXY, SILKY OR DULL
TRANSPARENCY	TRANSLUCENT TO OPAQUE
DISPERSION	NONE
BIREFRINGENCE	NONE
PLEOCHROISM	NONE

LOCATION China, England, Italy, New Zealand, Norway, Russia, United States.

▶ *Cabochon-cut serpentine gem.*

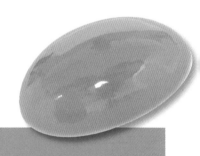

▶ *Serpentine crystal fragment.*

SERPENTINE

Most serpentine is a grayish-green color, and it gives this color to the rocks in which it occurs. However, it also occurs in other colors, which are popular for jewelry: **bowenite**, found in China, New Zealand, and the United States, is yellowish-green and resembles jadeite; **williamsite**, found in China, Italy, and the United States, is dark green.

PREHNITE

Until recent finds in China and New South Wales, Australia, prehnite was fairly rare. Other notable occurrences are in Asbestos, Québec, Canada; the Harz Mountains of Lower Saxony, northwest Germany; pale green masses in Stirling, Renfrew, and Dunbarton, Scotland; and aggregates of crystals exist in Alsace, northeast France. Other locations include the Black Forest, Baden Württemberg, southwest Germany; Pune, Maharashtra, western India; St Just and the Lizard Peninsula, Cornwall, southwest England; and the Isle of Mull, Argyll, western Scotland. US sites include New Haven County, Connecticut; Essex County and Middlesex County, Massachusetts; Keweenaw County, Michigan; and Passaic County, New Jersey. Pink and green specimens found in the Lake Superior area are prehnite and chlorite mixtures.

Baguette

Step

Cabochon

▲ *Prehnite crystals.*

Although it is hard enough to cut for jewelry, prehnite is mostly collected and is sometimes used as an ornamental stone. Its translucent property assures that it is cut into cabochons, some like cat's-eye, and also facet cut, although these gems are normally small. Step and baguette cuts are also popular. Prehnite is typically pale green to yellowish green with a white streak, but the color fades when extracted and exposed to air. It is named after a Dutchman, Colonel H. von Prehn (1733–85), who discovered the mineral at the Cape of Good Hope, South Africa, and brought the first specimen to Europe.

Prehnite occurs most commonly in veins and cavities in igneous rocks, often in association with zeolites. Associated minerals include QUARTZ, CALCITE, copper, stilbite, and datolite. Although harder, prehnite can be confused with SMITHSONITE, gyrolite, and hemimorphite. Prehnite occurs in very low-grade metamorphic rocks, and as a product of the decomposition of plagioclase feldspar. It can be an epimorph, which is a crystal growth over another mineral; a prehnite thick crust over laumontite is a striking example.

◄ *Cabochon-cut prehnite gem.*

PREHNITE	
CHEMICAL COMPOSITION	$Ca_2Al_2Si_3O_{10}(OH)_2$
COLOR	PALE GREEN, GRAY, YELLOW, WHITE
REFRACTIVE INDEX	1.61–1.64
RELATIVE DENSITY	2.9–3
HARDNESS	6–6.5
CRYSTAL GROUP	ORTHORHOMBIC
CLEAVAGE	BASIL, GOOD
FRACTURE	UNEVEN
TENACITY	BRITTLE
LUSTER	VITREOUS
TRANSPARENCY	TRANSPARENT TO TRANSLUCENT
DISPERSION	NONE
BIREFRINGENCE	.022–.033
PLEOCHROISM	NONE

LOCATION Australia, Austria, Canada, China, France, Germany, Italy, Namibia, Scotland, South Africa, United States.

QUARTZ

▲ *Carved rock crystal brooch with diamond and green tourmaline panel. The rock crystal has diamond palmette center clips.*

▼ *Rock-crystal and smoky-quartz crystals, amethyst and citrine, milky-quartz carving, star and rose quartz.*

The most abundant mineral, quartz is found in myriad colors, shapes, and varieties. It has been known and admired since antiquity, and its name derives from the ancient Greek *krustallos*, meaning "ice," because the Greeks (and the Romans) believed that quartz was ice that never melted because it was formed by the gods.

Quartz is formed in many different ways and occurs in many igneous and metamorphic rocks, particularly in granite and gneiss, and it is abundant in clastic sediments. It is also a common gauge mineral in mineral veins, and most good crystals are found in this type of occurrence. Well-formed quartz crystals can be obtained from cavities (geodes), from granite porphyries and from granite pegmatites. Because it resists weathering, quartz is also found in alluvial sands and gravels.

There are two main varieties of quartz: crystalline quartz, which is discussed here, and cryptocrystalline quartz, in which the crystals are so small as to be microscopic. This latter group includes amongst others CHALCEDONY, AGATE, JASPER, ONYX, and CARNELIAN.

What is remarkable about crystalline quartz is its amazing range of crystal size and color. It occurs in a range of colors, from clear ROCK CRYSTAL to pink ROSE QUARTZ or purple AMETHYST, to dark SMOKY QUARTZ. Sometimes minerals trapped within the crystals create beautiful effects, as with CHATOYANT QUARTZ or RUTILATED QUARTZ. The crystals can be huge, up to 22 feet (7 meters) in circumference and weighing 50 to 70 tons, or they can be delicate and beautifully formed.

For the collector, this infinite variation in form is the appeal of quartz. So is its cost – as quartz is so abundant, it is generally quite affordable. Many varieties make attractive jewelry, and quartz is also tough enough to be carved and shaped into many different shapes and ornaments. Quartz vibrates at a steady rate in reaction to an electrical charge and synthetic quartz is used in watches to keep time.

ROCK CRYSTAL

Brilliant

Step

Bead

Cabochon

For thousands of years, rock crystal has been associated with divination and healing. Famously, it has been used for crystal balls, although today it is rare to find crystals of sufficient size and clarity. The ancient Romans, who believed that rock crystal was permanently frozen ice, kept crystal balls in their atria, resting their hands on the cool stone on hot days.

The most common variety of QUARTZ, rock crystal is colorless and transparent. It is widely distributed. It has a beauty of its own, which, combined with the fact that it is easily cut, means that it is used as a gemstone in its own right. Although it lacks fire and brilliance, it is also sometimes a substitute for DIAMOND. It is also used for carved ornaments, especially on chandeliers, and has numerous industrial applications.

Quartz occurs in many igneous and metamorphic rocks, especially in granite and gneiss. It is abundant in clastic sediments. Quartz is also a common gauge mineral in mineral veins, and most good rock crystals are found in this type of occurrence. Fine rock crystals can be obtained from geodes and from granite porphyries and pegmatites. Because it is fairly resistant to weathering, rock quartz is also present in gravel and alluvial sands.

Rock crystal often has inclusions. Often these can add to the appearance and value of the specimen, as with rutilated quartz. Sometimes, a crystal can be found with inclusions of water and carbon dioxide: known as a two-phase inclusion, this appeals to collectors because the movement of the gas within the water can sometimes be seen with the naked eye.

▲ *Cut rock crystal gem.*

▶ *Crystals of rock crystal.*

ROCK CRYSTAL	
CHEMICAL COMPOSITION	SiO_2
COLOR	COLORLESS
REFRACTIVE INDEX	1.54–1.55
RELATIVE DENSITY	2.6–2.65
HARDNESS	7
CRYSTAL GROUP	TRIGONAL
CLEAVAGE	WEAK IN THREE DIRECTIONS
FRACTURE	CONCHOIDAL
TENACITY	TOUGH
LUSTER	VITREOUS
TRANSPARENCY	TRANSPARENT TO TRANSLUCENT
DISPERSION	LOW, .03
BIREFRINGENCE	.009
PLEOCHROISM	NONE
LOCATION Worldwide.	

ROCK CRYSTAL

Collectors are intrigued by the various forms that rock crystal can take. Often it grows out of a matrix of MILKY QUARTZ, and its tall and slender crystals, rising from a single base, are very attractive. The best crystals come from Hot Springs, Arkansas; Cumbria, England; St Gotthard, Switzerland; and Brazil and Madagascar.

▲ *Rectangular-cut amethyst, pavé-set pink sapphire, and diamond ring.*

Boasting the famed color of royalty, amethyst sets the color standard for all other purple gemstones. A variety of quartz, amethyst has long been prized and has been included in royal collections from ancient Egypt to the British crown jewels.

Amethyst, like other quartz varieties, is created in many different ways and occurs in many igneous and metamorphic rocks, particularly in granite and gneiss. It is abundant in clastic sediments. It is also a common gauge mineral in mineral veins, and most good crystals are found in this type of occurrence. Some of the best amethyst is found in cavities (geodes) from granite porphyries and pegmatites. Because it resists weathering, amethyst is also found in alluvial sands and gravels.

Amethyst is interesting in that it varies greatly from one location to another, and experts can identify the source mine of a specimen based purely on a visual inspection. For example, while amethyst from Veracruz, eastern Mexico, tends to be very pale and the crystals are usually "phantomed" with clear quartz on the interior and purple on the outside, specimens from Guerrero, southwest-central Mexico, are "phantomed" the other way around and are some of the most valuable amethyst crystals.

The color of amethyst is unstable and can diminish with protracted exposure to sunlight. Amethyst can also be heat-treated to produce the yellow of the rarer quartz variety, CITRINE. Pale stones may be set in a closed setting with a backing of foil to enhance the color.

The name amethyst derives from the ancient Greek word *amethustos* meaning "sober," and it was said that amethyst could prevent the bearer from becoming drunk, which was why wine goblets were sometimes made of the gem. In Greek mythology, amethyst was rock crystal dyed purple by the tears of Dionysus, the god of wine and revelry. In fact, it owes its purple color to impurities of iron.

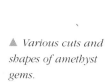

▲ *Various cuts and shapes of amethyst gems.*

▶ *Amethyst crystals on matrix.*

Today, amethyst is very popular for jewelry. It is found in both long prismatic crystals, which are suitable for cutting, or as druses. Because color can be patchy, it is often cut as round brilliants to maximize color. Deep colors are the most valuable.

Brilliant

▲ *Mixed-cut gem.*

Amethyst is graded, with the best-quality, darkest specimens judged "Siberian," no matter what their actual source; mid-quality stones are called "Uruguayan"; and lesser-quality specimens are "Bahain." Pale amethyst is sometimes called "Rose de France." Poor-quality stones are often tumbled to make beads or are cut as cabochons.

Sometimes amethyst is part of a mixed crystal: when it alternates with colorless quartz, it is known as amethyst quartz; when it bands with citrine, it is called ametrine.

Baguette

Bead

Mixed

AMETHYST

Today, the best examples of amethyst come from Jalgaon, Maharashtra, western India; Rio Grande do Sol, southern Brazil; and Ratnapura, southwest Sri Lanka. Other sources are variable. In general, South American crystals are larger than African, but the African amethyst has more saturated color. "Uruguayan amethyst" is famed for flashes of red.

◀ *Necklace with 11 tumbled amethyst beads spaced by pavé-set yellow diamond boules, and joined by a rhodochrosite clasp.*

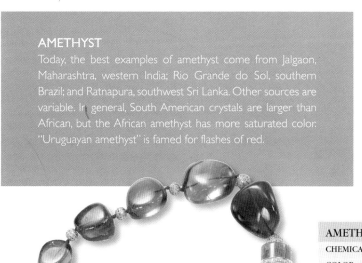

AMETHYST	
CHEMICAL COMPOSITION	SiO_2
COLOR	PURPLE
REFRACTIVE INDEX	1.54–1.55
RELATIVE DENSITY	2.65
HARDNESS	7
CRYSTAL GROUP	TRIGONAL
CLEAVAGE	NONE
FRACTURE	CONCHOIDAL
TENACITY	TOUGH
LUSTER	VITREOUS
TRANSPARENCY	TRANSPARENT TO TRANSLUCENT
DISPERSION	LOW, .03
BIREFRINGENCE	.009
PLEOCHROISM	NONE

LOCATION Australia, Brazil, Canada, Germany, India, Madagascar, Mexico, South Africa, Sri Lanka, Uruguay, United States.

Brilliant

Cabochon

Pendeloque

Natural, bright yellow citrine is the rarest of the QUARTZ varieties. Named for the old French word *citrin*, meaning "lemon," it has been popular for thousands of years and revered for its rarity. The ancient Romans used it for jewelry and intaglio work, and it was very popular for jewelry in the 19th century.

Citrine occurs in igneous and metamorphic rocks, particularly in granite and gneiss. It is also found in clastic sediments. Most good crystals are found as gauge minerals in mineral veins. Well-formed quartz crystals can be obtained from geodes, and from granite porphyries and pegmatites. Citrine is often found in association with amethyst, but it is much rarer than its purple cousin. Because it resists weathering, citrine is also found in alluvial sands and gravels.

Today, most citrine is artificially created, as amethyst turns yellow when heat-treated. It is difficult to distinguish natural from artificially created citrine, although heat treatment does tend to produce a slightly red tint. Because citrine is heat-sensitive, specimens should be protected from excessive exposure to heat or light. When citrine and amethyst combine in a banded stone, the gem is known as ametrine and is also popular.

Citrine is sometimes confused with TOPAZ, which is much more precious and expensive. Be wary of stones called "topaz quartz," or "citrine topaz," or the like, because they signal the fact that the less costly citrine is being passed off as something more valuable.

▲ *Carved citrine gem from Brazil, 12.80 carats.*
◄ *Large citrine crystal.*

CITRINE

By far the largest supplier of natural citrine is Rio Grande do Sol state in southern Brazil. Citrine mines in the US are located in North Carolina, Colorado, and California. The most valuable stones are the darkest, sometimes known as "Madeira citrine" for their resemblance to the color of fortified wine. In nature, citrine's color is due to tiny impurities of ferrous oxide.

CITRINE	
CHEMICAL COMPOSITION	SiO$_2$
COLOR	YELLOW TO AMBER
REFRACTIVE INDEX	1.54–1.55
RELATIVE DENSITY	2.6–2.7
HARDNESS	7
CRYSTAL GROUP	TRIGONAL
CLEAVAGE	WEAK IN THREE DIRECTIONS
FRACTURE	CONCHOIDAL
TENACITY	TOUGH
LUSTER	VITREOUS
TRANSPARENCY	TRANSPARENT TO TRANSLUCENT
DISPERSION	LOW, .03
BIREFRINGENCE	.009
PLEOCHROISM	NONE
LOCATION Brazil, France, Madagascar, Russia, United States.	

ROSE QUARTZ

▲ *Rose quartz, sapphire and enamel clock by Cartier.*

With a color unique among gems, rose quartz is probably the most popular of the quartz varieties. Its pink hues have adorned ornaments and jewelry since ancient times.

Most rose quartz is found in the cores of pegmatites. It is nearly always found in massive or lump form – crystals are extremely rare and so far have only been discovered in Minas Gerais, southeast Brazil. This is puzzling because quartz generally is known for its abundant crystals.

The cause of rose quartz's pink tint is thought to be due to tiny impurities of manganese, titanium or iron. Care must be taken, because it is heat sensitive and can fade to gray with prolonged exposure to air or sunlight.

Like other forms of quartz, rose quartz is prone to inclusions. Rutile needles can create asterism, but even here rose quartz is unlike other gems. While most asterisms appear when light is shone on a stone, with rose quartz the asterism is only visible when light is viewed through the stone.

Most rose quartz is too cloudy to be faceted and instead is cut as a cabochon. Transparent specimens, which are so pale as to be almost colorless, are very rare and valuable. In addition to its use in jewelry, rose quartz is easily carved into a wide range of ornaments. When purchasing rose quartz, be aware that pale specimens are sometimes dyed to create a more vivid color.

Bead

Mixed

Cameo

ROSE QUARTZ

While most rose quartz is mined in Minas Gerais, especially Araçuaí, the best-quality stones come from Samiresy, Antsirabe, central Madagascar. Other sources include Vama, Nuristan, north-east Afghanistan; Quadeville, Renfrew County, Ontario, Canada; and the Bavarian Forest of southern Germany. Major US sites include Fremont County, Colorado; Oxford County, Connecticut; and Grafton, New Hampshire.

ROSE QUARTZ	
CHEMICAL COMPOSITION	SiO_2
COLOR	PINK
REFRACTIVE INDEX	1.54–1.55
RELATIVE DENSITY	2.65
HARDNESS	7
CRYSTAL GROUP	TRIGONAL
CLEAVAGE	NONE
FRACTURE	CONCHOIDAL
TENACITY	TOUGH
LUSTER	VITREOUS
TRANSPARENCY	GENERALLY TRANSLUCENT TO OPAQUE, RARELY TRANSPARENT
DISPERSION	MEDIUM, .013
BIREFRINGENCE	.009
PLEOCHROISM	WEAK TO DISTINCT

LOCATION Brazil, Germany, India, Madagascar, United States.

▼ *Flower-cut rose-quartz gem from Brazil, 2.80 carats.*

▲ *Crystal fragment of rose quartz.*

Mixed

Cameo

Smoky quartz is one of the rare brown gemstones – the others being smoky TOPAZ, TOURMALINE, BERYL, and brown CORUNDUM. Like other varieties of quartz, it has been used and appreciated since ancient times because it is easy to cut as a gem and equally easy to shape for ornaments and practical applications.

Smoky quartz occurs in many igneous and metamorphic rocks, particularly in granite and gneiss. It is often found in quartz veins and granite pegmatic dykes. It also often occurs in granitic rocks that have small amounts of radioactivity.

While the cause of its color is not known for certain, it is believed that smoky quartz is brown to gray because it has been exposed to radiation. In fact, it can be made in the laboratory by exposing clear ROCK CRYSTAL to low-grade radiation (this effect can be reversed by heating). Today, most of the smoky quartz on the market has been color enhanced or modified in some way. One indication of heat treatment is an excessive uniformity of color, particularly in very dark specimens. In crystals, natural smoky quartz is brown to the base, while irradiated specimens will show some white next to the matrix rock.

SMOKY QUARTZ

Smoky quartz sometimes has different names based on its provenance and coloration. Black specimens are known as **morion**; crystals and gems from Scotland's Cairngorm Mountains are called **cairngorm**; and crystals that show black and gray banding are named **raccoon-tail quartz.**

Sources include Victoria, Australia; Rio Doce, Minas Gerais, southeast Brazil; Monteagle Township, Hastings County, Ontario, Canada; Haut-Rhin, Alsace, northeast France; Fichtelgebirge, Bavaria, southern Germany; Buryatia Republic, eastern Siberia, Russia. US sources include El Paso County, Colorado; Auburn, Maine; and Grafton County, New Hampshire.

◄ *Smoky quartz clock.*

▲ *Polished smoky quartz.*

► *Smoky quartz crystal.*

SMOKY QUARTZ

CHEMICAL COMPOSITION	SiO_2
COLOR	BROWN, GRAY OR BLACK
REFRACTIVE INDEX	1.54–1.55
RELATIVE DENSITY	2.65
HARDNESS	7
CRYSTAL GROUP	TRIGONAL
CLEAVAGE	WEAK IN THREE DIRECTIONS
FRACTURE	CONCHOIDAL
TENACITY	TOUGH
LUSTER	VITREOUS
TRANSPARENCY	TRANSPARENT TO TRANSLUCENT
DISPERSION	LOW, .03
BIREFRINGENCE	.009
PLEOCHROISM	DISTINCT

LOCATION Australia, Brazil, Germany, India, Madagascar, Russia, Switzerland, United Kingdom, United States.

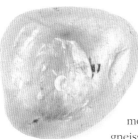

Brilliant

Cameo

Milky quartz owes its soft white color to microscopic inclusions of fluid that were trapped during its formation. While it is rarely faceted as a gemstone, it is found in attractive crystal formations and can be used for ornamental carvings.

Milky quartz occurs in many igneous and metamorphic rocks, particularly in granite and gneiss. It is abundant in clastic sediments. It is also a common gauge mineral in mineral veins, and most good crystals are found in this type of occurrence. Well-formed quartz crystals can be obtained from geodes, from granite porphyries, and from granite pegmatites. Because it resists weathering, milky quartz is also found in alluvial sands and gravels.

▲ *Rolled pebble of milky quartz.*

Milky quartz is interesting in the way in which it co-exists with other minerals. For example, it is often found as a "phantom" inside otherwise clear ROCK QUARTZ. It is also occasionally found associated with gold in hydrothermal veins, and the white stone with its gold inclusions is very attractive. When cut as a cabochon, milky quartz has an almost greasy luster and may be confused with opal.

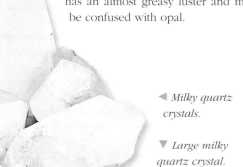

◄ *Milky quartz crystals.*

▼ *Large milky quartz crystal.*

MILKY QUARTZ	
CHEMICAL COMPOSITION	SiO_2
COLOR	WHITE
REFRACTIVE INDEX	1.54-1.55
RELATIVE DENSITY	2.65
HARDNESS	7
CRYSTAL GROUP	TRIGONAL
CLEAVAGE	NONE
FRACTURE	CONCHOIDAL
TENACITY	TOUGH
LUSTER	VITREOUS TO GREASY
TRANSPARENCY	TRANSLUCENT
DISPERSION	LOW, .03
BIREFRINGENCE	.009
PLEOCHROISM	DISTINCT

LOCATION Brazil, Madagascar, Namibia, Russia, United States.

MILKY QUARTZ

Sources of milky quartz include Bavaria, southern Germany; Plaka, Attica, east-central Greece; and the Great Wheal Fortune, Breal, Cornwall, southwest England. US locations include Moosup, Windham County, Connecticut; Grafton, Grafton County; New Hampshire; Westchester County, New York; and Providence County, Rhode Island.

Bead

Cabochon

Polished

Chatoyant quartz (or cat's-eye quartz) is found in a range of colors. The best known (and most common) is **tiger's-eye quartz**, a dark brown stone enlivened by yellowish fibers creating the cat's-eye effect. There are also two rarer varieties of cat's-eye quartz: a blue-green variety known as **hawk's-eye quartz**, and a greenish-gray variety also known as **cat's-eye quartz**. There is also a reddish brown version known as **bull's-eye quartz** or **ox-eye quartz**.

Cutting is crucial with chatoyant quartzes. In the rough, the stones reveal little or nothing of the exciting chatoyancy of the cut and polished stones. The gems are used for jewelry, small carvings such as cameos or intaglios, and ornaments.

Chatoyant quartz is often heat-treated. Most red ox-eye quartz is the result of heat treatment. Acid treatment produces a gray variation, and staining results in any number of colors. In addition, tiger's-eye quartz can be bleached to produce a stone similar in appearance to the more valuable cat's-eye CHRYSOBERYL.

▶ *Polished chatoyant quartz domes.*

CHATOYANT QUARTZ

Chatoyant quartz is a pseudomorph – the result of one mineral replacing another – of crocidolite, which occurs as veins in bedded ironstones. Crocidolite is sometimes known as blue asbestos, and its fibrous nature is what creates the long fibers in chatoyant quartz. In the case of hawk's-eye quartz, the crocidolite is permeated or partially replaced by quartz; tiger's-eye is the true pseudomorph, which has also undergone oxidation to produce the golden color. Chatoyant quartz is found in gravels and is also mined.

▶ *Polished and cabochon-cut chatoyant quartz.*

CHATOYANT QUARTZ	
CHEMICAL COMPOSITION	SiO$_2$
COLOR	BROWN, REDDISH-BROWN, GRAY
REFRACTIVE INDEX	1.54–1.55
RELATIVE DENSITY	2.46–2.71
HARDNESS	7
CRYSTAL GROUP	TRIGONAL
CLEAVAGE	NONE
FRACTURE	CONCHOIDAL OR SPLINTERY
TENACITY	TOUGH
LUSTER	VITREOUS TO SILKY
TRANSPARENCY	SEMITRANSLUCENT TO OPAQUE
DISPERSION	.013
BIREFRINGENCE	.0090
PLEOCHROISM	NONE
LOCATION Australia, Brazil, Burma (Myanmar), India, South Africa.	

RUTILATED QUARTZ

While most quartz is prized for its clarity, some inclusions create interesting and attractive effects. This is certainly the case for rutilated quartz, in which large inclusions of golden or red rutile needles are found in ROCK CRYSTAL. Rutilated quartz is used as a gemstone, cut as cabochons or even brilliants, and is also worked for ornamental carvings.

Inclusions such as rutile are incorporated into the quartz during the liquid stage of formation. Sometimes a secondary inclusion can create startling effects: for example, when HEMATITE is also present, the rutile needles will radiate out from the center of the hematite in a star shape, creating a gem known as **star rutile**.

Rutilated quartz has long captured the imagination and is known under a variety of names. Some are poetic, such as "Venus' hair" or *fleches d'amour* ("arrows of love"). **Needlestone** simply describes the appearance of the gem. It is also known as **sagenite**.

Other inclusions create similar effects in rock crystal. Tourmaline quartz is created when dark green or black TOURMALINE is trapped within clear quartz. GOLD and SILVER may also be included. Sometimes, the inclusion appears almost plantlike when green actinolite crystals are included, or dendritic quartz, when iron dendrites appear like ferns or little trees.

Brilliant

Bead

Cabochon

Cameo

▲ *Cabochon-cut rutilated quartz.*
▶ *Rutilated quartz crystal.*

RUTILATED QUARTZ

CHEMICAL COMPOSITION	SiO$_2$
COLOR	COLORLESS OR WHITE
REFRACTIVE INDEX	1.54–1.55
RELATIVE DENSITY	2.65
HARDNESS	7
CRYSTAL GROUP	TRIGONAL
CLEAVAGE	NONE
FRACTURE	CONCHOIDAL
TENACITY	TOUGH
LUSTER	VITREOUS
TRANSPARENCY	TRANSPARENT TO TRANSLUCENT
DISPERSION	LOW, .03
BIREFRINGENCE	.009
PLEOCHROISM	WEAK TO DISTINCT

LOCATION Australia, Brazil, Madagascar, South Africa, Sri Lanka, United States.

RUTILATED QUARTZ

Rutilated quartz is discovered in igneous and metamorphic rocks, particularly in granite and gneiss. It is abundant in clastic sediments. It is also a common gauge mineral in mineral veins, and most good crystals are found in this type of occurrence. Well-formed rutile quartz crystals can be obtained from geodes, from granite porphyries, and from granite pegmatites. Because it is fairly resistant to weathering, quartz is also found in alluvial sands and gravels.

CHALCEDONY

Chalcedony is the name given to compact varieties of silica which comprise minute, often fibrous, quartz crystals densely packed but with submicroscopic pores. The two main varieties are chalcedony, which is uniformly colored, and AGATE, characterized by curved bands or zones of differing color.

Chalcedony is precipitated from silica-bearing solutions and forms cavity linings, veins, and replacement masses in a variety of rocks. It is found in petrified wood and some other fossils. Chert and flint, which are impure forms of chalcedony, may originate by the deposition of silica on the sea floor or by the replacement of rocks, notably limestone, by silica from percolating waters.

Chalcedony's name is derived from Chalcedon or Calchedon, an ancient port of Bithynia, near present-day Istanbul, Turkey. The stone was much sought after in classical times as a gemstone and is still popular with jewelers and collectors as a gemstone and ornamental stone, often being artificially stained. Its popularity for carving and sculpture is due to its toughness and its occurrence in large, unbroken lumps.

Different colored varieties have individual names. CARNELIAN is red to reddish brown, SARD is light to dark brown, CHRYSOPRASE is apple-green, and BLOODSTONE, or heliotrope, is green with red spots. JASPER is generally red but sometimes yellow, brown, green, and gray-blue. **Moss agate** has a milky-white, bluish-white to nearly colorless matrix containing green and sometimes brown or black colors. Flint and chert are usually dull gray to black, and they fracture with sharp edges, a property that was exploited by early man to fashion implements.

Chalcedony occurs worldwide. Sources include Brazil, Uruguay, Sri Lanka, Russia, Iceland, Canada, Germany, Syria, Mexico, India, Madagascar, South Africa, Brazil, United Kingdom, and in several US states including California, Oregon, Florida, Arizona, Nevada, and Oregon.

▶ *Chalcedony has many different uses and displays many different colors depending on the specific cut.*

Jasper is a massive opaque variety that is a mixture of chalcedony, quartz, and opal. The name comes from the Latin name for the stone, *iaspis*, which probably also referred to other types of chalcedony. Jasper is found as small veins and replacements in different metamorphic and sedimentary rocks, and in cracks of volcanic rocks. It is fine-grained and colored by red and yellow iron oxides or green chlorite and actinolite. It often shows a combination of red, yellow, brown, green, and gray-blue. **Orbicular jasper** is a red variety containing white or gray patterns shaped like eyes. **Riband jasper** is striped, while **hornstone** is gray.

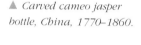

▲ *Carved cameo jasper bottle, China, 1770–1860.*

Jasper was used by the ancient Egyptians, Greeks, and Romans who carved portraits and such ornamental objects as amulets. It was believed that a person wearing jasper would be protected from illness, especially stomach problems. It has traditionally been used as a gemstone for jewelry such as brooches, earrings, necklaces, pendants, intaglios, and cameos, as well as for carvings and mosaics. It is normally cut as cabochons. Boulders of jasper weighing several hundred pounds (kilograms) have been sculpted in Ukraine and the Urals of Russia.

Cabochon

Polished

Cameo

JASPER	
CHEMICAL COMPOSITION	SiO_2
COLOR	RED, BROWN, YELLOW
REFRACTIVE INDEX	1.53–1.54
RELATIVE DENSITY	2.59–2.61
HARDNESS	6.5–7.0
CRYSTAL GROUP	TRIGONAL
CLEAVAGE	NONE
FRACTURE	CONCHOIDAL
TENACITY	TOUGH
LUSTER	VITREOUS TO WAXY
TRANSPARENCY	TRANSPARENT TO SUBTRANSLUCENT
DISPERSION	VERY WEAK
BIREFRINGENCE	.009
PLEOCHROISM	VERY WEAK

LOCATION Australia, Canada, Chile, Czech Republic, Egypt. Finland, France, Georgia, Germany, Iceland, India, Iran, Italy, Kazakhstan, Libya, Madagascar, Mexico, Norway, Paraguay, Russia, Slovak Republic, Sweden, Ukraine, United Kingdom, United States.

JASPER

Jasper occurs throughout the world. Various colors are found in the USA, such as orbicular jasper in California and jasperized fossil wood in Arizona. The Urals of Russia yield red, brown, green, and white riband jasper. Kazakhstan has a red-and-green variety, while Venezuela and India have red jasper. It also occurs in the Libyan desert, the Nile valley of Egypt, and the Rhineland of Germany.

▲ *Cabochon-cut jasper gem.*

▶ *Polished face of jasper boulder.*

Bead

Cabochon

Cameo

Carnelian (or cornelian) is a translucent chalcedony that receives its beautiful red tints from iron oxides, such as HEMATITE, embedded in the colorless silica. It may have either an unbroken color or be faintly banded. Its name comes from the Latin word *carneus*, which means "fleshy" – a reference to its color. The ancient Greeks and Romans particularly valued carnelian, which they mostly used for intaglios. It has

▲ *Cabochon-cut carnelian gem.*

also long been popular for signet rings. The Romans said that dark carnelian represented the male, while the light color symbolized the female. People also once believed that carnelian could calm bad temper or still the blood. Napoleon returned from his campaign in Egypt with an impressive octagonal carnelian stone.

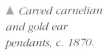

▲ *Carved carnelian and gold ear pendants, c. 1870.*

◀ *Orange and reddish carnelian polished slices.*

CARNELIAN

Most of today's commercial carnelian is stained chalcedony that comes from Brazil or Uruguay. It is also heat-treated to enhance its color. Beautiful stones come from India, where they are put in the sun to transform the brown color to red. Carnelian is a relatively inexpensive gemstone. Cut and polished as cabochons, it is also popular for beads and cameos.

The major sources of carnelian are India, Brazil, Uruguay, and Japan. Three blocks weighing more than 3 pounds (1.5 kilograms) have been discovered on the lower Narbada River, western India. Other locations include Queensland, Australia; Bohemia, Czech Republic; Franche-Comté, France; the Black Forest and Bavaria, Germany; Ratnapura, southwest Sri Lanka; Siberia, Russia; Cornwall, England; and the US states of Connecticut, Florida, New Jersey, Tennessee, and Washington.

CARNELIAN	
CHEMICAL COMPOSITION	SiO_2
COLOR	YELLOW, ORANGE, RED
REFRACTIVE INDEX	1.53–1.54
RELATIVE DENSITY	2.59–2.61
HARDNESS	6.5
CRYSTAL GROUP	TRIGONAL
CLEAVAGE	NONE
FRACTURE	CONCHOIDAL
TENACITY	TOUGH
LUSTER	VITREOUS TO WAXY
TRANSPARENCY	TRANSPARENT TO SUBTRANSLUCENT
DISPERSION	VERY WEAK
BIREFRINGENCE	VERY LOW, .009
PLEOCHROISM	VERY WEAK

LOCATION Australia, Brazil, Czech Republic, England, France, Germany, India, Japan, Madagascar, Russia, Slovak Republic, Ukraine, United States, Uruguay.

Bloodstone, also known as **heliotrope**, is an opaque green variety of chalcedony. The name was chosen because of the stone's many red spots (caused by HEMATITE), which resemble drops of blood. The ancient Greeks and Romans believed athletes would gain endurance by wearing it. In medieval times, some people believed the spots were originally left by Christ's blood splashing down from the cross, and the stones therefore had special powers, such as stopping a hemorrhage if touched. They also used the stone to carve sculptures depicting the crucifixion and martyrdom. The heliotrope name derives from the Greek words *helios*, meaning "sun," and *tropos*, meaning "turn," because the ancients believed it could turn the rays of the sun red and, if placed in water, show the sun as a blood-red image.

Bloodstone is porous and quite soft, but is a popular gemstone that is usually cut as a cabochon. It makes a popular choice for beads and pendants, as a sealstone (with an engraved device), and for men's signet rings. It has long been cut into cameos and used for decorative carvings. With heat treatment, the green background turns to gray and the red spots change to black. Another variety of chalcedony, closely related to bloodstone, is **plasma**, which is also green and may have yellowish spots. The name bloodstone sometimes is mistakenly used in reference to hematite and fancy JASPER.

Bead

Cameo

Polished

▲ *Two polished bloodstone cabochons.*

BLOODSTONE

CHEMICAL COMPOSITION	SiO_2
COLOR	DARK GREEN, GREENISH-BLUE WITH SMALL RED SPOTS
REFRACTIVE INDEX	1.53–1.54
RELATIVE DENSITY	2.59–2.61
HARDNESS	6.5
CRYSTAL GROUP	TRIGONAL
CLEAVAGE	NONE
FRACTURE	CONCHOIDAL
TENACITY	TOUGH
LUSTER	VITREOUS TO WAXY
TRANSPARENCY	TRANSPARENT TO SUBTRANSLUCENT
DISPERSION	VERY WEAK
BIREFRINGENCE	VERY LOW
PLEOCHROISM	VERY WEAK

LOCATION Austria, Australia, Brazil, China, Czech Republic, Germany, India, Italy, Russia, United States.

BLOODSTONE

The main source of bloodstone is the Kathiawar Peninsula, western India. It is also found in the Urals of Russia; the Harz Mountains in Germany; the Tyrol region of Austria and Italy; and Bohemia in the Czech Republic. Sources in the US include Wyoming, Maine, New York, California, Oregon, and Pennsylvania.

▶ *Polished bloodstone gem.*

Bead

Cabochon

Polished

Sard is similar to AGATE but has no banding. The name is derived from the Greek word *sardios* and Latin *sarda* for Sardis, the ancient capital of the kingdom of Lydia, Asia Minor (now Turkey), where the stone was originally found. Sard and CARNELIAN were first called sardion until medieval times, and this is the second oldest known name for a silica mineral (after crystal). Sard ("sardine stone") appears in the Exodus and Revelations books of the Christian Bible. It is one of the precious stones set into the breastplate of the high priest.

▲ *Polished oval slice of sard.*

The color differentiation between sard and carnelian is still confused, but sard gems are typically less intense in tone and more brown than carnelian. The stone's coloring is acquired by the presence of limonite, the hydrated iron oxide. Sard has been used for engraved jewelry since antiquity and was especially popular for such Victorian pieces as cameo rings, signet rings, and seals. It is an excellent material for inlay work, and is carved as cameos, intaglios, and small sculptures. It is also tumbled, cut and polished as beads, and also often polished as ovals.

Imitation sard is created by saturating chalcedony with an iron solution. Bands of sard and white ONYX are called SARDONYX, which at one time was more precious than gold, silver, or SAPPHIRE. Sardonyx is widely used in cameos and intaglios.

SARD	
CHEMICAL COMPOSITION	SiO_2
COLOR	BROWNISH RED
REFRACTIVE INDEX	1.53–1.54
RELATIVE DENSITY	2.59–2.61
HARDNESS	6.5
CRYSTAL GROUP	TRIGONAL
CLEAVAGE	NONE
FRACTURE	CONCHOIDAL
TENACITY	TOUGH
LUSTER	VITREOUS TO WAXY
TRANSPARENCY	TRANSPARENT TO SUBTRANSLUCENT
DISPERSION	VERY WEAK
BIREFRINGENCE	VERY LOW
PLEOCHROISM	VERY WEAK
LOCATION Australia, Brazil, China, India, Madagascar, Sri Lanka, United States.	

▶ *Sard on matrix.*

SARD

Sard occurs in Ratnapura, southwest Sri Lanka, and the Black Hills of Imperial County, California, United States. Other sources include India and China.

▲ *Polished slice of onyx.*

Onyx is a variety of AGATE, but instead of curved bands it has straight parallel ones of alternate black and white. If reddish-brown bands occur instead of black, it is called SARDONYX. The onyx name is generally applied in commercial dealings to the black bands, and this has led to any black chalcedony being called onyx, whether it is natural or dyed.

The ancient Romans carved different patterns in each layer of a multi-layered onyx and often used the stone for their seals, carving a design in negative relief to produce the raised print. They had first used the onyx name for a variety of marble having white and yellow veins, and *onyx* is the Greek word for "claw" or "fingernail," because these veins resemble the colors of a fingernail. The marble is still called "onyx marble," being less valuable and softer than onyx.

Onyx has been dyed since ancient times to enhance the color. It has traditionally been used for carving cameo brooches. Because of the stone's different bands, part of the black layer can be carved away to create a white background while leaving a black design on top. The colors can be reversed: Wedgewood ceramic ware, for instance, resembles onyx cameos with white figures over a dark gray background.

Onyx items are affordable, and include brooches, beads, earrings, necklaces, pendants, stones for rings, chess pieces, bookends, cups, vases, and figurines such as animals. The stone may scratch or chip easily. Natural black onyx is rare, so the commercial black variety is nearly always agate that has been stained by the sugar-sulfuric acid treatment: soaked in a sugar solution and then heated in sulfuric acid to carbonize the sugar.

Bead

Cabochon

Polished

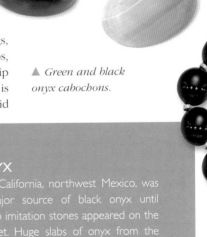

▲ *Green and black onyx cabochons.*

ONYX	
CHEMICAL COMPOSITION	SiO_2
COLOR	BLACK, DARK BROWN, GREEN
REFRACTIVE INDEX	1.53–1.54
RELATIVE DENSITY	2.59–2.61
HARDNESS	6.5
CRYSTAL GROUP	TRIGONAL
CLEAVAGE	NONE
FRACTURE	CONCHOIDAL
TENACITY	TOUGH
LUSTER	VITREOUS TO WAXY
TRANSPARENCY	TRANSPARENT TO SUBTRANSLUCENT
DISPERSION	VERY WEAK
BIREFRINGENCE	VERY LOW
PLEOCHROISM	VERY WEAK
LOCATION Afghanistan, Brazil, India, Madagascar, Mexico, Peru, United States.	

ONYX

Baja California, northwest Mexico, was a major source of black onyx until cheap imitation stones appeared on the market. Huge slabs of onyx from the open pit Onyx Mine at El Marmol are transported to nearby towns for cutting and polishing. Slabs have even been used to build the school house at El Marmol and graves are covered with onyx slabs.

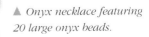

▲ *Onyx necklace featuring 20 large onyx beads.*

Bead

Cabochon

Polished

Cameo

Sardonyx is another AGATE variety with reddish-brown and white bands. Its name reflects its make-up, combining SARD and ONYX. The ancient Egyptians carved sardonyx into scarab beetles, wearing them as talismans. Roman soldiers went into battle believing they gained courage by wearing sardonyx engraved with Mars, the god of war. According to tradition, Queen Elizabeth I gave the Earl of Essex a ring of sardonyx set in gold as a keepsake. In Renaissance Europe, wearing a sardonyx stone was believed to bestow eloquence on speakers.

Much natural sardonyx exists, but a large amount is produced by staining agate, and it is sometimes called "banded agate." Sardonyx has long been used in engraved jewelry, and was once considered more valuable than silver or gold. It was carved for the oldest cameos known, and today is used for these and intaglios, because the layers can be cut to offset images in relief. The many other sardonyx items include beads, brooches, and small statuary. Relatively common and inexpensive, it is normally cut as cabochons and can be fashioned to resemble an eye.

▲ *Cabochon of sardonyx.*

SARDONYX

The best sardonyx has been mined in India for millennia. In the 2nd century, Egyptian geographer Ptolemy drew a map of India, which included "sardonyx stones" in the Sardonyx Mountains. India remains a major source. Other locations include the Ural Mountains of central Russia. Major localities in the United States include Tampa Bay, Florida, and El Paso County, Texas.

SARDONYX	
CHEMICAL COMPOSITION	SiO_2
COLOR	REDDISH BROWN WITH BLACK OR WHITE BANDS
REFRACTIVE INDEX	1.53–1.54
RELATIVE DENSITY	2.59–2.61
HARDNESS	6.5
CRYSTAL GROUP	TRIGONAL
CLEAVAGE	NONE
FRACTURE	CONCHOIDAL
TENACITY	TOUGH
LUSTER	VITREOUS TO WAXY
TRANSPARENCY	TRANSPARENT TO SUBTRANSLUCENT
DISPERSION	VERY WEAK
BIREFRINGENCE	VERY LOW
PLEOCHROISM	VERY WEAK
LOCATION	Australia, Brazil, Czech Republic, India, Madagascar, Russia, Sri Lanka, United States, Uruguay.

▶ *Cameo carved from sardonyx.*

Chrysoprase, also known as cat's-eye, is the most valued variety of chalcedony. The name comes from the Greek words *chrysos*, meaning "gold," and *prason*, meaning "leek," since it resembled the color of the vegetable. The color can vary from yellowish green to apple green and grass green, depending upon the content of hydrated silicates or nickel oxides. In transparency, it ranges from nearly opaque to nearly transparent. **Prase** is another form of green chalcedony. Found in eastern Europe and the US states of Delaware and Pennsylvania, it is very rare and has a less vivid green color caused by inclusions of actinolite; it is used for cameos, beads, and small statuary carvings. **Mtorolite** is the name given to a variety of green chalcedony colored by chromium and found in Zimbabwe.

The ancient Greeks and Romans valued chrysoprase as a decorative stone and for jewelry beads. It was very popular in the 14th century when Holy Roman Emperor Charles IV decorated chapels with chrysoprase, including the Chapel of St Wenceslas in Prague. The stone was also a favorite of Frederick the Great of Prussia and of Queen Anne of England.

Chrysoprase is cut as cabochons and its bright, even color and texture makes the stone ideal for beads and cameos. The color will eventually fade in heat and even sunlight, but it sometimes returns when chrysoprase is stored in high humidity. Excellent pieces have been sold as Imperial jade and Australian jade. Dyed agate and glass have also been sold as chrysoprase.

Bead

Cabochon

Cameo

CHRYSOPRASE	
CHEMICAL COMPOSITION	SiO$_2$
COLOR	YELLOWISH GREEN TO APPLE GREEN
REFRACTIVE INDEX	1.53–1.54
RELATIVE DENSITY	2.58–2.64
HARDNESS	6.5
CRYSTAL GROUP	TRIGONAL
CLEAVAGE	NONE
FRACTURE	CONCHOIDAL
TENACITY	TOUGH
LUSTER	VITREOUS TO WAXY
TRANSPARENCY	TRANSPARENT TO SUBTRANSLUCENT
DISPERSION	NONE
BIREFRINGENCE	.004
PLEOCHROISM	NONE
LOCATION Australia, Austria, Brazil, Czech Republic, Germany, Kazakhstan, Madagascar, Poland, Russia, South Africa, United States.	

CHRYSOPRASE

Early chrysoprase came from Bohemia, Czech Republic, but today the best and largest single source of chrysoprase is Marlborough in Queensland, Australia. Superb stones have long been mined in Lower Silesia, Poland. Other major locations include Saxony, east-central Germany; the Urals of central Russia; and Sarakul-Baldy, Kazakhstan. US sites include Gila County, Arizona, and Tulare County, California.

▲ *Silver girole (pendant) set with chrysoprase and pearls, designed (1903) by May Morris.*

▶ *Chrysoprase cabochons (below) and rough specimen (right).*

A gate is distinguished by its multiple colors that are often vivid. The bands of color may be white, red, blue, gray, brown, or black. These are irregular and sometimes curved, caused by traces of iron and manganese. The parallel bands follow the wavy contour of the cavity where silica gel solidified. If part of a cavity remains unfilled, crystals may form of AMETHYST, SMOKY QUARTZ, or ROCK CRYSTAL.

The stone's name was derived from the River Achates (now Dirillo), southwest Sicily, since it was originally discovered in that area. Agate is one of the oldest known minerals. It was valued by the ancient Sumerians and Egyptians who used it for amulets, receptacles, and ornamental pieces. The Greeks and Romans carved it into cameos and intaglios, and the latter used agate for intaglio signet rings. The stone was especially valued during medieval times. Today, it remains popular for brooches, cameos, and other jewelry, and as a decorative stone for inlaid work, such as boxes and table tops. In industry, it is used in the manufacture of grinding equipment.

▲ Polished agate slice (above), and a moss agate (right) and colorful agate cabochon (below).

Agates can be translucent, transparent, or sometimes opaque. Wavy bands of color are often inside a geode (nodule) of agate. The colorful bands are best seen when the geode is sawed through and the flat faces of the sections polished. Cutting the stone into thin slices allows light to pass through and display the bands. Many varieties are dyed or stained artificially, as was done in ancient Greece and Rome, and more expertly by artisans in Constantinople (now Istanbul, Turkey). Blue-stained agates are called "Swiss lapis" or "false lapis." The most expensive stones are rough specimens that are easily colored, especially

▼ Polished agate showing typical banding pattern.

AGATE

Agates form in cavities and cracks, mostly in volcanic rocks where they fill the cavities that were made by escaping gas during the solidification of molten rock. They form as minute fibers and crystals, instead of large crystals like quartz. Agate is found worldwide. Idar-Oberstein, Rhineland-Palatinate, western Germany was once the prime source, but this is now depleted. Today, excellent stones come from Rio Grande do Sul, southern Brazil, and fine naturally colored agate exists in Bohemia in the Czech Republic. Agate is also found in the Auvergne, south-central France; the Hansrück Mountains of Rhineland-Palatinate, western Germany and the Erzgebirge, eastern Germany; the Deccan region of southern India; the Urals of Russia; Salto, northwest Uruguay; and Montrose, Angus, eastern Scotland. Agate is a common mineral in much of North America. Major sites include British Columbia, Canada; San Bernadino County, California; and the western states of Idaho, Montana, Oregon, and Washington.

Cabochon

Cameo

Polished

Brazilian specimens which are more porous and have more opal content than other stones. All agates respond well to polishing, and they are strong enough for everyday jewelry.

There are many varieties of agate. **Plume agate** has a feather-like design. **Moss agate** has inclusions of chlorite that form greenish branched veins resembling a coating of moss on the milky-white background. **Dendritic agate** possesses a branching or tree-like design, while **red jasper agate** and **cloud agate** have dark centers in a cloudy white background. **Fortification agate** has angular, zigzag bands that suggest a fort, and **lace agate** is noted for its delicate patterns. **Turritella agate** is composed mostly of embedded turritella snail shells. **Iris agate** has iridescent colors reflected from between the layers. **Star agate** shows star-shaped designs, and **ruin agate** is a rare variety where movements of the Earth have broken up the filled cavities and it has re-formed in a different color. A gem from Mexico showing a single "eye" is called **cyclops agate**.

◀ *The Barbor Jewel (1590–1600) – an enameled gold pendant set with an agate cameo of Queen Elizabeth I.*

▲ *A 19th-century banded agate bowl from Khambhat (Cambay), Gujarat, western India.*

AGATE	
CHEMICAL COMPOSITION	SiO_2
COLOR	ALMOST ANY
REFRACTIVE INDEX	1.535–1.540
RELATIVE DENSITY	2.55–2.64
HARDNESS	6.5–7.0
CRYSTAL GROUP	TRIGONAL
CLEAVAGE	NONE
FRACTURE	CONCHOIDAL
TENACITY	TOUGH
LUSTER	VITREOUS TO WAXY
TRANSPARENCY	TRANSPARENT TO SUBTRANSLUCENT
DISPERSION	.013
BIREFRINGENCE	.005
PLEOCHROISM	WEAK TO DISTINCT

LOCATION Armenia, Australia, Brazil, Canada, Czech Republic, France, Georgia, Germany, India, Iran, Italy, Madagascar, Mongolia, Scotland, South Africa, Ukraine, United Kingdom, United States, Uruguay.

FIRE AGATE

Bead

Cabochon

Mixed

Cameo

Fire agate has a rusty brown color. When lapidaries grind and polish away layers, light passing through the inner layers of silica and iron oxide (limonite and goethite) creates iridescent colors known as "fire." The finest specimens have a resemblance to burning embers when properly cut. The red color is the most valued, and a variety with a mixture of blue and green is also popular, with brown being the least desired. The best pieces also have a continuous iridescent layering, although most specimens have these in patches.

The stone is relatively inexpensive, costing a fraction of the price of precious OPAL, which has a similar play of color. Cabochons are cut to highlight the rainbow color effect and mixed, round, and oval cuts are popular. The gem is used for such items as beads, pendants, and cameos. The hardness and durability make it a good choice for everyday jewelry like rings.

FIRE AGATE

Fire agate is not common or well known, having become commercially traded only in the late 1950s when many rough stones were mined in Mexico. Fire agate is also very difficult to cut, since the curvature of colored layers has to be respected, and the stone will be ruined if one layer too many is removed.

The stone occurs in Sonora, northwest Mexico, and in the southwest United States, such as Maricopa County, Arizona. It is formed as warm water saturated with silica and iron oxide fills cavities in rock and begins to cool.

FIRE AGATE	
CHEMICAL COMPOSITION	SiO_2
COLOR	RUSTY BROWN
REFRACTIVE INDEX	1.535–1.540
RELATIVE DENSITY	2.55–2.64
HARDNESS	6.5–7.0
CRYSTAL GROUP	TRIGONAL
CLEAVAGE	NONE
FRACTURE	CONCHOIDAL
TENACITY	TOUGH
LUSTER	VITREOUS TO WAXY
TRANSPARENCY	OPAQUE
DISPERSION	.013
BIREFRINGENCE	.005
PLEOCHROISM	WEAK TO DISTINCT
LOCATION Mexico, United States.	

◄ *Polished slab of fire agate.*

► *Crafted fire agate crystal.*

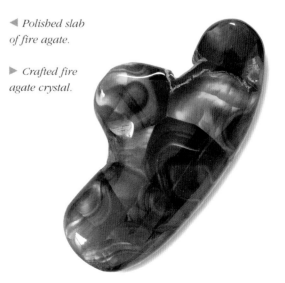

Step

Cabochon

Cameo

Opal is a hydrated silica, which is a hardened gel of silica and water. The name comes from the Sanskrit word *upala* or Greek *opalios*, which means "gem" or "precious stone." The refraction of light by tightly packed silica spheres in precious opal creates a play of color, or "fire," within the gem when it is turned. **Common opal**, also known as "potch," is translucent or sometimes opaque, having a range of colors but lacking the fire of precious opal. It is therefore of little or no value.

▲ *Polished opal from Mexico.*

Opals were prized as much as diamonds by the ancient Greeks and used for jewelry by the Romans. In the 1st century, Roman scholar and naturalist Pliny the Elder (AD 23–79) wrote that opal "has an unbelievable mixture of colors." The Aztecs of Mexico mined fine pieces that were taken back to Europe by the Spanish *conquistadors*.

Opal is amorphous, having no crystal structure. (The only other major amorphous gemstone is AMBER.) It comes in massive, stalactitic, and rounded forms. It is deposited at low temperatures from silica-bearing waters and contains up to 10% water. If the stone dries out too quickly after being mined, it may crack. Opal can occur as a fissure-filling in rocks of any kind, but especially in areas of geysers and hot springs. It can be formed during the weathering and decomposition of rocks. Diatomite, or diatomaceous earth, is a fine-grained sedimentary rock made largely of opaline silica skeletons of organisms such as sponges and diatoms. It is used as an abrasive, insulator, filler, and filtration powder.

Fire opal is a variety in which red and yellow colors are dominant and produce flame-like reflections when turned. **Hyalite** is colorless, botryoidal opal. **Wood opal** is wood that has been replaced in part by opaline silica. **Geyserite** (or siliceous sinter) is an opal deposited by geysers and

▲ *Fire opal gem.*

▲ *Polished boulder opal.*

▶ *Opal crystal on matrix.*

OPAL

The most valuable opals come from Queensland and New South Wales in Australia, which has been the main producer since their discovery in the 1870s. Fire opal is mined in Hidalgo, southern Mexico, and is also found in Kazakhstan and Honduras. Romania has variously colored deposits in Transylvania. In the US, precious opal exists in Nevada, diatomite in California, and wood opal in Yellowstone National Park, Wyoming. Other sites include Arizona, Colorado, Idaho, Oregon, Washington, and Georgia. Wood opal is also found on Sardinia, Italy.

▶ *Group of rough and worked specimens and items of opal jewelry.*

▼ *Cabochons of opal gems in various colors.*

hot springs, and **hydrophane** is a variety which becomes transparent when immersed in water.

The color of opals is generally determined by metallic oxides and foreign inclusions. Exceptionally fine **black opals** can be as expensive as diamonds – Lightning Ridge in New South Wales is a famous source. Reddish fire opals are also prized. The other sought-after varieties are **white opal**, which has a good play of color, and colorless **water opal** which is clear with flashes of color. The best-quality stones are transparent. Many opals are cut as cabochons and others as step cuts. Fire opals are faceted. Opal is popular for jewelry, but is soft and can be easily damaged. It is also carved as cameos. Imitation opals have been made using **Slocum stone**, a man-made glass that gives a play of color. Chips of opal and colored plastic are also put into hollowed rock crystal, and an imitation opal from Gilson laboratory uses silica spheres.

OPAL	
CHEMICAL COMPOSITION	$SiO_2.nH_2O$
COLOR	COLORLESS, MILKY WHITE, GRAY, RED, BROWN, BLUE, GREEN, NEARLY BLACK
REFRACTIVE INDEX	1.43–1.46
RELATIVE DENSITY	1.8–2.3
HARDNESS	5.5–6.5
CRYSTAL GROUP	AMORPHOUS
CLEAVAGE	NONE
FRACTURE	CONCHOIDAL
TENACITY	BRITTLE
LUSTER	VITREOUS TO RESINOUS, SOMETIMES PEARLY
TRANSPARENCY	TRANSPARENT TO SUBTRANSLUCENT
DISPERSION	NONE
BIREFRINGENCE	NONE
PLEOCHROISM	NONE
LOCATION Australia, Brazil, Czech Republic, Guatemala, Honduras, Italy, Mexico, Romania, South Africa, United States, Zimbabwe.	

▶ *Wood opal.*

FELDSPAR

A component of most kinds of rocks, feldspar is an almost ubiquitous mineral, making up around 60% of the Earth's crust. But for all that it is so commonplace, of the nearly 40 members of the feldspar group so far identified, only around nine are familiar.

Although they tend to crystallize in igneous environments, feldpars are also present in many metamorphic rocks. The name feldspar refers to crystalline aluminosilicate minerals, which are divided into two groups depending on their chemical composition. The **potassium** types consist of ORTHOCLASE, MICROCLINE, and others; and the **plagioclase** types, in which sodium (Na) and calcium (Ca) substitute for each other in a continuous series, include ALBITE, LABRADORITE, and OLIGO-CLASE (gem variety, SUNSTONE).

Despite this chemical variation, all feldspars have certain characteristics in common. Twinning is frequent. Feldspars have no color of their own, relying on impurities to add coloration. They all also have two good cleavages, of which one is perfect. At this perfect cleavage, the luster is pearly, and in cracks perpendicular to it, iridescent colors may be reflected.

Cleavage properties are at the root of the mineral's name. It is named for the German *feld*, meaning "field," and *spar*, meaning "easily cleaved mineral." In fact, feldspars are widely distributed and are often found in topsoil, where they release plant nutrients into the soil as they weather.

Feldspar is frequently put to industrial use – in the manufacture of glass, plumbing fixtures, tiles, and pottery. But some varieties are also prized as gemstones. Many are admired for their special sheen, or schiller – while LABRADORITE has the best iridescence, MOONSTONE is the most valuable and also best known of the gem varieties.

▼ *Rough and worked specimens of albite, oligoclase and labradorite – all examples of plagioclase feldspar.*

Brilliant

Step

Mixed

Of the potassium feldspar types, it is the orthoclase varieties that are most often used in jewelry. Appealing for their clarity, orthoclase are also relatively inexpensive. Their tendency to twin also produces many crystals that attract the eye of collectors. Although twinning is common, perfect twins in a single crystal are very rare and highly sought after.

Orthoclase is the common potassic feldspar of most igneous and metamorphic rocks. It tends to be found in granites and syenites that cooled moderately quickly. It is also found as grains in sedimentary rocks. Sanidine is more common in extrusive igneous rocks, like volcanic rocks, that cooled very quickly. It also occurs in rocks that have been thermally metamophosed at high temperature.

The orthoclase family gets its name from its cleavage properties. It comes from the Greek *orthos*, meaning "straight," and *kalo*, meaning "cleave" or "break," because the cleavage planes are virtually perpendicular. Several varieties of orthoclase are used as gemstones. The most popular is known as **adularia**. It is a clear and colorless variety that benefits from a bluish-white sheen, known generally as "schiller" or sometimes "adularescence." Because of this iridescent sheen, adularia is often confused with MOONSTONE. However, while moonstone is translucent, adularia is clear or semitransparent. It was very popular in Art Nouveau jewelry in the early 20th century. Sanidine is another clear form of orthoclase that is sometimes gem-quality. The yellow orthoclase, deriving its color from impurities of iron, is also faceted, usually into step cuts because it tends to be somewhat fragile. Yellow orthoclase is easily confused with LABRADORITE, a more valuable member of the feldspar family. The two can readily be distinguished by their specific gravities – labradorite is the denser of the two. Its type locality is the Adula Mountains, St Gotthard, Ticino, southern Switzerland.

ORTHOCLASE

CHEMICAL COMPOSITION	$K(Al,Si)_4O_8$
COLOR	OFF-WHITE, YELLOW, PINK TO RED, ORANGE TO BROWN
REFRACTIVE INDEX	1.51–1.54
RELATIVE DENSITY	2.55–2.58
HARDNESS	6–6.5
CRYSTAL GROUP	MONOCLINIC
CLEAVAGE	GOOD IN TWO DIRECTIONS
FRACTURE	CONCHOIDAL TO UNEVEN
TENACITY	BRITTLE
LUSTER	DULL TO VITREOUS
TRANSPARENCY	TRANSLUCENT TO OPAQUE
DISPERSION	VERY LOW, .012
BIREFRINGENCE	WEAK TO MEDIUM, .005–.015
PLEOCHROISM	NONE

LOCATION Australia, Burma (Myanmar), Germany, India, Italy, Madagascar, Mexico, Sri Lanka, Switzerland, United States.

▶ *Oval-cut orthoclase.*

◀ *Orthoclase crystals on matrix.*

ORTHOCLASE

Found in Sri Lanka and Burma (Myanmar) as well as Germany, it facets nicely into step and table cuts. The best specimens come from Madagascar, although stones from Germany, when cut as cabochons, sometimes display a pleasing chatoyancy (cat's-eye).

Long considered a sacred stone in India, moonstone is often believed to be moonlight in a magical, solid form. A virtually colorless stone, it is distinguished by a bluish shimmer that moves across its face as the stone is moved.

Moonstone's magical shimmer, often known as schiller, is caused by the intergrowth of ORTHOCLASE and ALBITE, another feldspar variety. To show it at its best, the gem is generally cut as a cabochon, and the higher the dome, the more radiant the shimmer. It is also sometimes carved into ornaments. Moon faces are a popular theme, but over the years it has even been used in cameos. Moonstone's shimmer is distinctively blue. If the sheen displays in a variety of hues, the stone is known as "rainbow moonstone," which is actually a variety of LABRADORITE, another type of feldspar found in India and other countries.

Popular in the Art Nouveau jewelry of the early 20th century, today moonstone is again in demand. It is rare – and becoming rarer – which makes it reasonably expensive. Because it is expensive, moonstone is sometimes imitated, using either glass or decolorized AMETHYST. In addition to its beauty, moonstone is believed to possess strong protective and healing powers.

Cushion

Cabochon

Cameo

▲ Moonstone and opal ear pendants. The moonstones are at top and bottom.

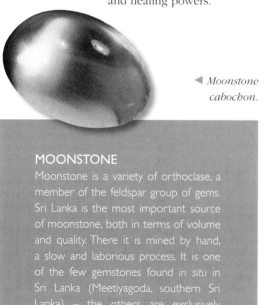

◀ Moonstone cabochon.

MOONSTONE

MOONSTONE	
CHEMICAL COMPOSITION	$K(Al,Si)_4O_8$
COLOR	COMMONLY COLORLESS OR PALE
REFRACTIVE INDEX	1.51–1.57
RELATIVE DENSITY	2.56–2.62
HARDNESS	6–6.5
CRYSTAL GROUP	MONOCLINIC
CLEAVAGE	PERFECT IN THREE DIRECTIONS
FRACTURE	CONCHOIDAL TO UNEVEN
TENACITY	BRITTLE
LUSTER	DULL TO VITREOUS
TRANSPARENCY	TRANSPARENT TO TRANSLUCENT
DISPERSION	VERY LOW, .012
BIREFRINGENCE	WEAK TO MEDIUM, .005–.015
PLEOCHROISM	NONE

LOCATION Australia, Brazil, Burma (Myanmar), India, Madagascar, Mexico, Norway, Sri Lanka, United States.

MOONSTONE

Moonstone is a variety of orthoclase, a member of the feldspar group of gems. Sri Lanka is the most important source of moonstone, both in terms of volume and quality. There it is mined by hand, a slow and laborious process. It is one of the few gemstones found *in situ* in Sri Lanka (Meetiyagoda, southern Sri Lanka) – the others are exclusively found in gravel.

▼ Moonstone crystal fragment.

Bead

Cabochon

Cameo

The bluish-green color of **amazonite**, the gemstone variety of microcline, is often confused with pale JADEITE or lesser-quality TURQUOISE. It is an attractive stone in its own right, famously used by the ancient Egyptians for both jewelry and ornamentation.

Microcline is so called in reference to its cleavage properties. The name derives from the Greek *micros*, meaning "small," and *klino*, meaning "lean," because the cleavage planes deviate very slightly from 90 degrees.

For collectors, microcline is interesting because of its crystal habits. It can form the largest known crystals of any mineral, and large specimens come to market reasonably often. They are generally opaque, and translucent crystals are more valuable. Microcline often forms crystals in association with QUARTZ, known as graphic granite. When the match is between gem-quality amazonite and SMOKY QUARTZ, the resulting specimen is very striking.

Amazonite gets its green color from small impurities of lead. For jewelry it is usually cut *en cabochon*. It is frequently veined with white or yellow bands of plagioclase, which can be attractive.

MICROCLINE

CHEMICAL COMPOSITION	$K(Al,Si)_4O_8$
COLOR	COLORLESS, WHITE, YELLOW, GREEN, BLUE
REFRACTIVE INDEX	1.52–1.53
RELATIVE DENSITY	2.54–2.57
HARDNESS	6–6.5
CRYSTAL GROUP	TRICLINIC
CLEAVAGE	PERFECT IN ONE DIRECTION, GOOD IN ONE DIRECTION
FRACTURE	CONCHOIDAL
TENACITY	BRITTLE
LUSTER	VITREOUS
TRANSPARENCY	TRANSLUCENT TO OPAQUE
DISPERSION	VERY LOW, .012
BIREFRINGENCE	.008
PLEOCHROISM	NONE

LOCATION Australia, Brazil, Madagascar, Mozambique, Namibia, Russia, United States, Zimbabwe.

▶ *Microcline crystals on quartz matrix.*

MICROCLINE

A polymorph of ORTHOCLASE but with a triclinic crystal structure, microcline is another of the feldspar minerals. Microcline is formed in deep-seated igneous rocks that cooled slowly. It is found in granites, granite pegmatites, hydro-thermal veins, and in many schists and gneisses. It is also found as grains in sedimentary rocks.

Amazonite's name is a bit of a misnomer. It is named after the River Amazon, but although the stone is mined in Brazil, it has never been found near the Amazon. Brazilian legend has it that Amazon Indians used to give a green stone to men who met them, but that stone was probably JADEITE, not amazonite.

◀ *Cabochon-cut amazonite gem.*

ALBITE

A white or colorless stone, albite, one of the FELDSPAR group of minerals, is best known to collectors as a partner to rarer and more colorful gemstones.

The name feldspar refers to crystalline aluminosilicate minerals, which are divided into two groups depending on their chemical composition. Albite is the sodium end member of the plagioclase series.

Although it is not that interesting on its own, albite is often found in association with some rare minerals, since rare elements tend to get isolated in the later stages of crystallization. In this way, albite is often seen as an accessory mineral with TOURMALINE, SPODUMENE, or BERYL.

Albite can also form in layers with potassium feldspar minerals, especially ORTHOCLASE and MICROCLINE. The albite is visible as white stripes. When this occurs, the resultant stone is known as **perthite**. Albite content is also responsible for the special shimmer (schiller) of MOONSTONE, also a feldspar variety.

Brilliant

◀ *Albite crystals on matrix.*

▲ *Two brilliant-cut colorless albite gems.*

ALBITE

Albite is found in pegmatites and in sodic lavas. It is the last of the feldspars to crystallize from the molten rock.

The type locality for albite is Finnbo, Falun, Dalarna, Sweden. It is a quartz quarry in a pegmatite. Fluorite, tourmaline, topaz, apatite, beryl, and cassiterite are also found associated with the pegmatite.

▲ *White albite crystals on matrix.*

ALBITE	
CHEMICAL COMPOSITION	$Na(Al,Si)_4O_8$
COLOR	USUALLY WHITE, SOMETIMES GREEN
REFRACTIVE INDEX	1.53
RELATIVE DENSITY	2.6
HARDNESS	6–6.5
CRYSTAL GROUP	TRICLINIC
CLEAVAGE	PERFECT IN ONE DIRECTION, GOOD IN ONE DIRECTION
FRACTURE	CONCHOIDAL TO UNEVEN
TENACITY	BRITTLE
LUSTER	DULL TO VITREOUS
TRANSPARENCY	TRANSLUCENT TO OPAQUE
DISPERSION	VERY LOW, .012
BIREFRINGENCE	.010
PLEOCHROISM	NONE
LOCATION	Canada, Germany, Norway.

LABRADORITE

Cabochon

Polished

An Inuit legend has it that the Northern Lights were once imprisoned in some rocks on the Canadian coast, until a brave warrior freed most of them with a blow from his spear. Those that remained in the rock are the cause of the special iridescence of labradorite, a variety of feldspar. Its unique sheen means that labradorite is rarely mistaken for any other gem.

The name feldspar refers to crystalline aluminosilicate minerals, which are divided into two groups depending on their chemical composition. The potassium types consist of ORTHOCLASE, MICROCLINE, and others; and the plagioclase types include ALBITE, labradorite, and OLIGOCLASE – labradorite falls midway through the series. Labradorite, found in igneous and metamorphic rocks, is characteristic of gabbros and basalts. Rarely found as crystals, it is usually uncovered as a compact aggregate.

▲ *Labradorite cabochon.*

Many feldspar minerals display eye-catching optical effects, known as schiller. MOONSTONE is probably the best-known example. But while moonstone's special gleam is always bluish green, the schiller of labradorite may be blue, violet, green, yellow or orange, or even all of these at once (when this is the case, the resulting gem is known as **rainbow moonstone**). This phenomenon, known as labradorescence, is caused by the play of light on lamellar growths inside the stone that mean that light is reflected back and forth internally. A rough stone may show some glints of labradorescence, but it takes polishing to bring out a specimen's true colors.

Today, although it is very abundant, it is not often seen because it is difficult to cut, requiring special skill to bring out its colors. It is best cut as a low cabochon.

LABRADORITE

CHEMICAL COMPOSITION	Ca[50–70%] Na[50–30%]$(Al,Si)_4O_8$
COLOR	GRAY TO SMOKY BLACK
REFRACTIVE INDEX	1.55–1.57
RELATIVE DENSITY	2.70–2.72
HARDNESS	6–6.5
CRYSTAL GROUP	TRICLINIC
CLEAVAGE	PERFECT IN ONE DIRECTION, GOOD IN ONE DIRECTION
FRACTURE	CONCHOIDAL TO UNEVEN
TENACITY	BRITTLE
LUSTER	DULL TO VITREOUS
TRANSPARENCY	TRANSPARENT TO TRANSLUCENT, OCCASIONALLY OPAQUE
DISPERSION	VERY LOW, .012
BIREFRINGENCE	WEAK TO MEDIUM, .005–.015
PLEOCHROISM	NONE

LOCATION Australia, Canada, Costa Rica, Finland, Germany, Madagascar, Mexico, Norway, Russia, United States.

► *Labradorite crystal – when polished it appears to change from gray to blue.*

LABRADORITE

Named for Labrador, Newfoundland, northeast Canada, where the gem was first identified by geologists in the 1770s, it was popular in jewelry in France and England in the 18th century. The best stones still come from Labrador, but Finland, where it is known as **spectrolite**, is also an important source.

OLIGOCLASE (SUNSTONE)

With its bright, metallic glitter and its reddish to orange color, sunstone, the gemstone variety of oligoclase feldspar, seems to radiate warming rays. It has a long history – it has been found in Viking burial mounds, and historians believe that it was valued as a navigational aid, and Native American tribes used it in their Medicine Circles, trusting in its healing powers.

Feldspars are crystalline aluminosilicate minerals, which divide into two groups depending on their chemical composition: the potassium types, and the plagioclase types, which include ALBITE, LABRADORITE, and oligoclase, with oligoclase at the sodium-rich end of the series. Although they tend to crystallize in igneous environments, they are also present in many metamorphic rocks. Oligoclase is frequently formed in lava flows.

Sunstone is evaluated based on its depth of color and the orientation of its schiller. It is usually cut as low cabochons. Rarely, transparent specimens are found that can be faceted, and these are very valuable. The Norwegian sunstone is considered the best in the world, although the Indian stones, with their medium to dark orange color, are also appreciated. The Oregon sunstone, although new to the market, is already well respected.

Sunstone goes under a variety of names, including **aventurine feldspar** and **heliotite**. While it has no rivals among naturally occurring gemstones, beware of goldstone, a man-made imitation created from glass and copper.

Cabochon

▲ *Sunstone cabochon set into a hairpin.*

OLIGOCLASE (SUNSTONE)

CHEMICAL COMPOSITION	$Na[90-70\%]$ $Ca[10-30\%](Al,Si)_4O_8$
COLOR	OFF-WHITE OR GRAY, PALE YELLOW, GREEN, BROWN, RED TO ORANGE
REFRACTIVE INDEX	1.53–1.55
RELATIVE DENSITY	2.62–2.65
HARDNESS	6–6.5
CRYSTAL GROUP	TRICLINIC
CLEAVAGE	PERFECT IN ONE DIRECTION, GOOD IN ONE DIRECTION
FRACTURE	CONCHOIDAL
TENACITY	BRITTLE
LUSTER	DULL TO VITREOUS
TRANSPARENCY	TRANSPARENT TO TRANSLUCENT
DISPERSION	NONE
BIREFRINGENCE	.008
PLEOCHROISM	NONE
LOCATION	Canada, India, Norway, Russia, United States.

OLIGOCLASE (SUNSTONE)

Until the 1800s, when good deposits were found in Siberia, sunstone was a rare and very expensive gemstone. But even then, supply was erratic and the stone was little known. In the late 1990s, reserves were discovered in the northwest US state of Oregon, and today sunstone is a popular and widely recognized gem. While most sunstone gets its schiller from impurities of HEMATITE and goethite, Oregon sunstone glimmers because of inclusions of copper.

▶ *Oval cabochon of sunstone.*

SODALITE

Cabochon

Polished

▶ *Sodalite cabochon with white calcite from Namibia, 2.56 carats.*

This rich blue stone takes its name from its high sodium content. It is a constituent of the gem LAPIS LAZULI, with which it is sometimes confused; however, it lacks the sparkle of lapis, which is caused by the presence of brassy colored pyrite specks.

Sodalite is a major member of the feldspathoid group of minerals. These minerals are similar to FELDSPARS but with less silica content. They are formed in alkali-rich but silica-poor environments, as in alkaline igneous rocks such as nepheline syenites, and also in some silica-poor dyke rocks and lavas. For example, sodalite is found on Mount Vesuvius, Naples, southern Italy.

There are only three large deposits of sodalite in the world. The most important source is the Bancroft Mine (sometimes known as the "Princess Mine"), Dungannon Township, Hastings County, Ontario, eastern Canada. Other major deposits are in the Ice River complex, British Columbia, western Canada, and around Litchfield, Kennebec County, Maine, northeastern United States.

Sodalite's pure blue color makes it a popular stone for jewelry and ornaments. Sometimes called the "wisdom stone," it is said to enhance understanding and concentration, and also to bolster courage. Attractive samples often contain stripes of white CALCITE. It is generally cut flat or as a cabochon.

SODALITE

Clear crystals are very rare and, although some may be sourced either from Mount Vesuvius in Italy, or from the Aris Quarry, Windhoek, central Namibia, they are rarely large enough to be faceted.

Other locations of sodalite include Tasmania, Australia; Ayopaya province, Cochabamba, western Colombia; the Eifel Mountains, Rhineland-Palatinate, western Germany; and the Kola Peninsula, northern Russia.

SODALITE	
CHEMICAL COMPOSITION	$Na_8Al_6Si_6O_{24}C_{12}$
COLOR	COMMONLY BLUE, ALSO PINK, YELLOW, GREEN, OR GRAY-WHITE
REFRACTIVE INDEX	1.48
RELATIVE DENSITY	2.1–2.3
HARDNESS	5–6
CRYSTAL GROUP	CUBIC
CLEAVAGE	RHOBDODECAHEDRAL, POOR
FRACTURE	CONCHOIDAL TO UNEVEN
TENACITY	BRITTLE
LUSTER	VITREOUS
TRANSPARENCY	TRANSPARENT TO TRANSLUCENT
DISPERSION	NONE
BIREFRINGENCE	NONE (ISOTROPIC)
PLEOCHROISM	NONE

LOCATION Australia, Brazil, Canada, Colombia, Germany, Greenland, Italy, Namibia, Russia, United States.

◀ *Rough specimen of sodalite.*

Brilliant

A rare silicate mineral, haüyne is noted for its intense blue color. It was first discovered in 1807 among Vesuvian lavas on Monte Somma in southern Italy and was named after R. J. Haüy (1743– 1822), a French mineralogist and pioneer in crystallography. Haüyne occurs typically as small, rounded grains in silica-poor, alkali-rich lavas such as phonolites composed of alkali feldspars and nepheline. Volcanoes in Germany and Morocco are well-known sources. The mineral also occurs in LAPIS LAZULI, being similar in composition to lazurite, as well as other members of its sodalite group, such as SODALITE, nosean, bicchulite, kamaishilite, tugtupite, and tsaregorodtsevite.

▲ *Haüyne cabochon of 0.71 carats.*

Cabochon

Haüyne rarely occurs as individual crystals but is grown together with other minerals. Twinning is common in these well-formed coarse crystals. Cutting is rare and difficult because of the perfect cleavage, so pieces are generally faceted for collectors. The gems are priced high, especially those specimens with a deep sapphire-blue color.

▶ *The deep blue grains of haüyne are clearly visible.*

▶ *Blue haüyne on matrix.*

HAÜYNE

Other than Monte Somma of Vesuvius, deposits are found in the Lazio and Tuscany regions of west-central Italy. Other major sites include Cygnet, Tasmania, Australia; Catalonia, northeast Spain; the Eifel Mountains, Rhineland-Palatinate, western Germany, and Kaiserstuhl, Baden Württemberg, southwest Germany; and the remote Sumaco Volcano, north-central Ecuador. US deposits of haüyne include Sussex County, New Jersey; Colfax County, New Mexico; and St Lawrence County, New York.

HAÜYNE

CHEMICAL COMPOSITION	$Na_6Ca_2Al_6Si_6O_{24}(SO_4)_2$
COLOR	BLUE, ALSO GRAY, BROWN, YELLOW-GREEN
REFRACTIVE INDEX	1.50
RELATIVE DENSITY	2.4–2.5
HARDNESS	5.5–6
CRYSTAL GROUP	CUBIC
CLEAVAGE	PERFECT
FRACTURE	UNEVEN
TENACITY	BRITTLE
LUSTER	VITREOUS TO GREASY
TRANSPARENCY	TRANSPARENT TO TRANSLUCENT
DISPERSION	————
BIREFRINGENCE	NONE (ISOTROPIC)
PLEOCHROISM	NONE

LOCATION Australia, Germany, Italy, Morocco, Spain, Sweden, United States.

Lapis lazuli is a rock, a contact metamorphosed limestone that contains lazurite. It also contains PYRITE, which adds a golden-yellow sparkle, and CALCITE, which shows as white flecks. Other trace minerals can include HAÜYNE, SODALITE, DIOPSIDE, wollastonite, amphibole, feldspar, mica, APATITE, SPHENE, ZIRCON, and pyroxenes. The stone was once powdered and mixed with oil to produce the pigment ultramarine (literally, "beyond the sea"), which is seen in the beautiful blues of Renaissance paintings. Ultramarine has been made synthetically since 1828. *Lapis* is Latin for "stone." The names of both lazuli and lazurite are derived from the Persian word *lazhuward* and Latin word *lazulum*, meaning "blue" or "heaven." The lapis lazuli name, often shortened to lapis, is sometimes mistakenly used for the mineral lazurite.

Lapis lazuli was mentioned in 2650 BC in the Sumerian epic of Gilgamesh. The ancient Egyptians used it extensively in religious ceremonies, and lapis items were found in the tomb of Tutankhamen. It was a popular stone in Mesopotamia, Persia, and the ancient city of Ur, which had a large trade in lapis lazuli. The Greeks and Romans used it as a reward for bravery. The ancients also employed it for inlaid work and for jewelry, amulets, and talismans. They named it *sapphirus* ("blue"), which is now used for the blue corundum variety of SAPPHIRE. Lapis lazuli is mentioned in the Biblical book of Exodus. In the 17th century, it was used in medicine to prevent miscarriages, epilepsy, and dementia.

▶ *Oval lapis lazuli at the center of a brooch pendant with turquoise, carnelian, and diamonds.*

▲ *Rough specimen of lapis lazuli.*

▼ *Blue crystal of lapis lazuli.*

LAPIS LAZULI

Lapis lazuli usually occurs in crystalline limestone and is formed in contact metamorphic rocks associated with pyrite and calcite. It normally appears in massive and compact forms or as fine, granular aggregates. The best lapis lazuli is found in limestone in the Kokcha river valley of Badakhshan province in northeastern Afghanistan, and these deposits in the mines of Sar-e-Sang have been worked for more than 6,000 years. The Arab geographer Istakhri recorded a visit there in the 10th century, and Marco Polo visited and wrote about the lapis mines in c. 1271.

Other occurrences include light blue boulders at the western end of Lake Baikal, southern Siberia, Russia; pale stones in the Andes, near Ovalle, central Chile; and a dark variety in the Rockies of Colorado, USA. It is also found in Italy, near Rome and on Monte Somma, Vesuvius. Other sources include Baffin Island, Nunavut, Canada; San Bernardino County, California; and the Pamir Mountains of central Asia.

Cabochon

Cameo

Polished

The value of lapis lazuli is largely determined by the abundance and color of the dark intense blue lazurite. The colors range from greenish-blue to purple-blue. The flecks of gold pyrite and white calcite can often increase the value or, when too numerous or too large, decrease it. Lapis lazuli is cut and polished to make gemstones for jewelry and is also used as a decorative stone. Because it is slightly soft, it is normally cut as cabochons, and also used for beads, inlay material, and small carved items. The opaque mineral takes a good polish but may lose its luster because of its softness. A Gilson laboratory imitation was produced in the mid 1970s but it is more porous. Other imitations are made using the gem JASPER artificially stained blue, known as "Swiss lapis," and synthetic SPINEL colored blue by cobalt.

LAPIS LAZULI	
CHEMICAL COMPOSITION	$(Na,Ca)_{7-8}(Al,Si)_{12}$ $(O,S)_{24}(SO_4Cl,OH)_2$
COLOR	AZURE-BLUE, TRANSLUCENT
REFRACTIVE INDEX	1.50
RELATIVE DENSITY	2.4
HARDNESS	5.0–5.5
CRYSTAL GROUP	CUBIC
CLEAVAGE	POOR
FRACTURE	UNEVEN
TENACITY	
LUSTER	VITREOUS
TRANSPARENCY	TRANSLUCENT
DISPERSION	NONE
BIREFRINGENCE	NONE
PLEOCHROISM	NONE

LOCATION Afghanistan, Angola, Burma (Myanmar), Canada, Chile, Italy, Pakistan, Russia, Tajikistan, United States.

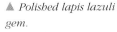

▲ *Polished lapis lazuli gem.*

◀ *Three different cuts of lapis lazuli.*

SCAPOLITE

Brilliant

Step

Cabochon

▲ *Violet fancy-cut scapolite gem.*

Scapolite refers to a series of gemstones ranging in color from clear to orange to purple because of their varying chemical compositions, from sodium-rich **marialite** to calcium-rich **meionite**. It is commonly found as prismatic crystals, giving rise to its name – scapolite comes from a combination of two Greek words, *scapos* for "rod," and *lithos* for "stone." It is also sometimes known as **wernerite**, after the German geologist A. G. Werner (1749–1817).

Scapolite is interesting to collectors because it belongs to a very rare symmetry class: the tetragonal dipyramidal class. Only powellite and SCHEELITE are also members of this class. This means that it has only one primary four-fold axis of rotation and a perpendicular mirror plane, but unfortunately it is rare to find crystals in which this can be seen. The best samples are prized for their color, particularly when yellow to orange or pink to violet. Only a few gems more than 15 carats have been found. When cut, the stones often exhibit a cat's-eye effect, but because they easily splinter they are difficult to fashion. Their low hardness means that they are not suitable for use in rings, and the stones must be treated with care.

▶ *Scapolite crystals on matrix.*

SCAPOLITE

CHEMICAL COMPOSITION	$(Na,Ca)_4(Si,Al)O_{24}(Cl,CO_3,SO_4)_{12}$
COLOR	CLEAR TO WHITE OR BLUISH GRAY, OR YELLOW TO ORANGE AND PINK TO VIOLET
REFRACTIVE INDEX	1.54–6.0
RELATIVE DENSITY	2.56–2.77
HARDNESS	5.5–6.0
CRYSTAL GROUP	TETRAGONAL DIPYRAMIDAL
CLEAVAGE	PRISMATIC IN TWO SETS, GOOD
FRACTURE	SUBCONCHOIDAL
TENACITY	BRITTLE
LUSTER	VITREOUS
TRANSPARENCY	TRANSLUCENT TO TRANSPARENT
DISPERSION	MEDIUM (0.17)
BIREFRINGENCE	.012–0.36
PLEOCHROISM	DISTINCT

LOCATION Australia, Burma (Myanmar), Canada, Madagascar, Mexico, Norway, Sweden, Tajikistan, Tanzania, United States.

▶ *Scapolite cut as a round cabochon.*

SCAPOLITE

Scapolite occurs in metamorphic rocks, particularly in metamorphosed limestone. It is also found in skarns close to igneous contacts and as a replacement of feldspars in altered igneous rocks.

First found in Burma (Myanmar), today the best-quality scapolite gemstones come from the Umba River area of northeast Tanzania.

Acommon sodium zeolite mineral, colorless and transparent natrolite has icy clear radiating sprays and is popular with collectors. There are also varieties that are white or slightly tinted yellow or brown. It was confirmed as a distinct species and named in 1803. The name comes from the Greek words *natron* meaning "soda," in reference to its sodium content, and *lithos* for "stone." Natrolite is fashioned into small faceted gems, such as table cuts, but the demand for this is not great. Massive material for cutting is especially mined in Scotland and Vestfold, southern Norway.

Table

Natrolite occurs typically in the cavities of basaltic and other igneous rocks. Like most zeolites, it has a porous structure of well-defined channels on an atomic scale that makes it into a molecular sieve. Water and large ions can travel throughout the crystal structure. Natrolite is often found with two other fibrous zeolites: mesolite and scolecite. The three are closely related and easily confused in the field because they have a similar acicular (needle-like) habit. However, natrolite is orthorhombic and the pseudo-orthorhombic mesolite and pseudo-tetragonal scolecite are both monoclinic. Natrolite's other associated minerals include benitoite, apophyllite, QUARTZ, RHODOCHROSITE, neptunite, and analcime.

◀ *Table-cut natrolite gem.*

NATROLITE

Notable occurrences of natrolite, besides Scotland and Norway, include Victoria, Australia; Pune, western India; Bohemia in the Czech Republic; Ice River Complex, British Columbia, western Canada, and Île de Montréal County, Québec, eastern Canada; Bavaria, southern Germany; Veszprém County, western Hungary; Sardinia, Italy; and the Kola Peninsula, northern Russia. US sites include San Benito County, California; Horseshoe Dam, Arizona; Passaic County, New Jersey; and Kings Mountain, North Carolina.

NATROLITE	
CHEMICAL COMPOSITION	$Na_2Al_2Si_3O_{10}2H_2O$
COLOR	COLORLESS, WHITE, YELLOWISH, BROWNISH
REFRACTIVE INDEX	1.48–1.49
RELATIVE DENSITY	2.2–2.3
HARDNESS	5–5.5
CRYSTAL GROUP	ORTHORHOMBIC
CLEAVAGE	PERFECT IN TWO DIRECTIONS
FRACTURE	UNEVEN
TENACITY	BRITTLE
LUSTER	VITREOUS
TRANSPARENCY	TRANSPARENT TO TRANSLUCENT
DISPERSION	NONE
BIREFRINGENCE	.0038–.004
PLEOCHROISM	NONE

LOCATION Australia, Canada, Czech Republic, Germany, Hungary, India, Italy, Norway, Russia, Scotland, Sweden, United States.

▶ *Natrolite crystals on matrix.*

THOMSONITE

Bead

Cabochon

Thomsonite is a rare member of the zeolite group, occurring in cavities in lavas, and as a decomposition product of nepheline. Because it has open channels, the mineral acts as a chemical sieve, allowing some ions to pass through but blocking others. It was named in 1820 after Thomas Thomson (1773–1852), a Scottish chemist who discovered it in the Kilpatrick Hills of Dunbartonshire, west-central Scotland. It is similar to NATROLITE but is usually more coarsely crystalline. Thomsonite is also difficult to distinguish by sight from other fibrous zeolites. Associated minerals also include CALCITE, QUARTZ, stilbite, and calchabazite.

Thomsonite crystals have an acicular (needle-shaped) form in radiating or divergent aggregates. Fibrous, spherical, and tabular crystals also occur. Gemstones are normally pure white, but collectors prize those tinged with red, green, yellow, and purple. These crystals have the luster of porcelain, and this is improved by faceting. Thomsonite is rarely cut as a gem. Cabochons are cut for rings, and thomsonite beads are popular.

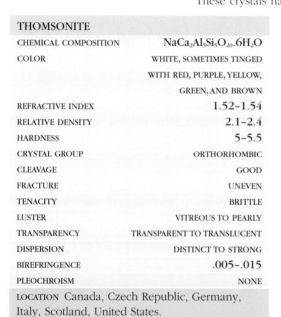

▶ *Thomsonite on matrix.*

THOMSONITE

CHEMICAL COMPOSITION	$NaCa_2Al_5Si_5O_{20}.6H_2O$
COLOR	WHITE, SOMETIMES TINGED WITH RED, PURPLE, YELLOW, GREEN, AND BROWN
REFRACTIVE INDEX	1.52-1.54
RELATIVE DENSITY	2.1-2.4
HARDNESS	5-5.5
CRYSTAL GROUP	ORTHORHOMBIC
CLEAVAGE	GOOD
FRACTURE	UNEVEN
TENACITY	BRITTLE
LUSTER	VITREOUS TO PEARLY
TRANSPARENCY	TRANSPARENT TO TRANSLUCENT
DISPERSION	DISTINCT TO STRONG
BIREFRINGENCE	.005-.015
PLEOCHROISM	NONE

LOCATION Canada, Czech Republic, Germany, Italy, Scotland, United States.

THOMSONITE

Sources of thomsonite include Flinders, Victoria, Australia; Severočeský Kraj, Bohemia, Czech Republic; the Eifel Mountains, Rhineland-Palatinate, western Germany; Sardinia, Italy; Kaipara, North Island, New Zealand; Kola Peninsula, northern Russia; and Invernesshire, northern Scotland. Gem-quality pieces in the US are found along the shores of Lake Michigan and Lake Superior, especially Keweenaw County, Michigan. Other US deposits include Grant County, Lane County and Wheeler County, all in Oregon.

▶ *Cuff links made from cabochons of thomsonite with cabochon rubies.*

◀ *Acicular (needle-like) crystals of thomsonite.*

Obsidian is the best-known glassy rock. It is a volcanic glass and is supposedly named after Obsidian, a Roman said to have brought the first specimens from Lake Shalla, Ethiopia, to Rome. The rock originates from explosive volcanoes. It is made of the same minerals as granite but cooled too quickly to crystallize. Obsidian has no crystalline structure but may contain rare tiny crystals of QUARTZ, FELDSPARS, pyroxene or magnetite.

Being a siliceous glass, it breaks with a conchoidal fracture producing sharp slivers that must be handled carefully. Obsidian may be fashioned into a razor-sharp cutting edge, and ancient civilizations used it for jewelry, mirrors, arrowheads, spearheads, scrapers, and cutting tools, such as the sacrificial knives of the Aztecs.

Today, transparent specimens are faceted, usually into step cuts, while less transparent pieces are fashioned into cabochons or table cuts. Brilliants are also popular. Dark nodules found in Mexico and in Arizona and New Mexico in the United States are called "**Apache tears**," which Native Americans have long used for decorative pieces and jewelry, such as necklaces, bracelets, and amulets. Also especially prized as jewelry is the rare **snowflake obsidian**, with white patches that are internal bubbles or crystals of potassium feldspar. Another favorite is the silky lustered variety with a sheen caused by minute crystals ("crystallites").

Step

Cabochon

Polished

OBSIDIAN

CHEMICAL COMPOSITION	SiO_2 (+ IMPURITIES)
COLOR	BLACK, SOMETIMES BROWN, GRAY, REDDISH
REFRACTIVE INDEX	1.48–1.53
RELATIVE DENSITY	2.33–2.60
HARDNESS	5.5–7
CRYSTAL GROUP	AMORPHOUS
CLEAVAGE	NONE
FRACTURE	CONCHOIDAL
TENACITY	BRITTLE
LUSTER	VITREOUS
TRANSPARENCY	TRANSLUCENT TO SEMITRANSPARENT OR OPAQUE
DISPERSION	VERY LOW
BIREFRINGENCE	NONE
PLEOCHROISM	NONE

LOCATION Armenia, Ecuador, Germany Guatemala, Hungary, Iceland, Indonesia, Italy, Japan, Mexico, New Zealand, Russia, Scotland, Slovak Republic, United Kingdom, United States.

OBSIDIAN

Occurrences in the USA include the Obsidian Cliffs in west-central Oregon, the Obsidian Cliff at Yellowstone National Park in Wyoming, the Glass Buttes in Oregon (entirely of obsidian), and other states, including New Mexico, Utah, and Hawaii. Other locations include Cumbria in England; the Lipari Islands of Italy; and Mexico, Iceland, Japan, Java in Indonesia, Hungary, Slovak Republic, Ecuador, and Guatemala.

▲ *Polished cabochon and heart-shaped obsidian.*

▶ *Obsidian with typical conchoidal fracture.*

TEKTITE (MOLDAVITE)

Tektites are small glassy objects that, unlike meteorites, are found only in certain, rather limited areas of the Earth's surface. Chemically, tektites comprise a silica-rich glass which is also rich in aluminum, potash, and lime, and can be matched by a few igneous and sedimentary rocks. The name comes from the Greek word *tektos*, meaning "melted." Tektites are usually small, the majority being less than 10 ounces (300 grams) in weight, and about 0.5 to 1 inch (1 to 3 centimeters) across, but some examples up to 25 pounds (12 kilograms) are recorded. The shape of tektites is variable, but discoid, lensoid, button-shaped, tear-drop, dumb-bell, spherical, and boat shapes commonly occur.

▶ *Marquise and heart-shaped faceted moldavite.*

Tektites are mostly crustal in composition with uneven or rough surfaces that show cracks from cooling. This has led to theories for their origin being the melting of terrestrial rocks through the impact of large meteorites or comets with the Earth. The **australite** button form lent strong support to the theory that tektites come from outside the atmosphere, since they melted on the forward side and this flowed toward the back. This heat causes part of a meteorite and Earth rock to melt and combine in a molten mass, like a glass meteorite that is sometimes known as a "melt droplet." Some contain bubbles that are round or shaped like a torpedo, and there are no crystalline inclusions as seen in volcanic obsidian. Other theories proposed an extraterrestrial origin, with meteors melting as they passed through the atmosphere, but a terrestrial origin is now favored.

TEKTITE (MOLDAVITE)

In 1787, the first tektites were discovered in the River Moldau (the German name for River Vltava) in Bohemia, Czechoslovakia – hence their original name of **moldavites**. Tektites began to be discovered in other countries around 1860, and the moldavite name is now mostly restricted to specimens from the Czech Republic and finds in Germany and Austria. Tektites are named according to the area in which they are found, and the other principal types are **australites** from the southern part of Australia, Tasmania (sometimes called "Darwin glass"), and coastal islands; **philippinites** from the Philippines and southern China; **javaites** from Java, Indonesia; **malaysianites** from Malaysia; **indochinites** from Thailand, Burma (Myanmar), China, Laos, and Vietnam; **Ivory Coast tektites** from the Ivory Coast; **bediasites**, first discovered at Bedias, Grimes County, Texas, USA; and **Georgia tektites** from Georgia, USA. Libyan desert glass is sometimes listed as a tektite but is more likely to be an impactite. Volcanic glass specimens from Peru and Colombia were once throught to be tektites and were named **americanites**. It has been estimated that something like 650,000 tektites have been collected, of which the **philippinites** account for some 500,000.

australites

philippinites

javaites

malaysianites

indochinites

Ivory Coast tektites

bediasites

Georgia tektites

moldavites

▶ *Various shapes of tektites.*

Brilliant

Cushion

Bead

Some tektites are smooth and shiny but others have a rough, strongly etched, and eroded surface, often with a system of grooves which reflect flow patterns within the glass. Most tektites are jet-black but thin flakes are transparent or translucent in shades of brown. The moldavites, however, are dark green and in thin flakes are transparent and bottle-green. They have an intense vitreous luster and are most suitable for faceting, and these are cut and polished as beads and fashioned into brilliant and cushion cuts for jewelry. Uncut moldavites are more valuable and are set in gold and silver. Specimens have also been carved into small, decorative items.

TEKTITE (MOLDAVITE)	
CHEMICAL COMPOSITION	$SiO_2,Al_2O_3,Fe_2O_3,$ FeO,MgO,CaO,Na_2O,K_2O
COLOR	USUALLY DARK GREEN
REFRACTIVE INDEX	1.48-1.50
RELATIVE DENSITY	2.3-2.4
HARDNESS	5.5-6.5
CRYSTAL GROUP	AMORPHOUS
CLEAVAGE	NONE
FRACTURE	VARIOUS
TENACITY	BRITTLE
LUSTER	VITREOUS
TRANSPARENCY	TRANSPARENT
DISPERSION	NONE
BIREFRINGENCE	NONE
PLEOCHROISM	NONE

LOCATION Australia, Austria, China, Czech Republic, Germany, Indonesia, Ivory Coast, Libya, Malaysia, Philippines, Thailand, United States, Vietnam.

▶ *Mottled green tektite.*

◀ *Map of the major finds of tektites.*

223

SYNTHETICS & IMITATIONS

▲ *Cubic zirconia (CZ) gem.*

A synthetic gemstone is one that is artifically created. Jewelers sometimes draw a distinction, saying a **simulated gem** imitates a real stone (like colored glass used for ruby), but a synthetic gem is a real gem produced in a laboratory and having the same chemical composition, atomic structure, and physical properties of the natural stone. Synthetic diamond can be produced by mimicking the diamond's natural growth using a high-pressure, high-temperature synthesis process or a chemical vapor deposition (CVD) with a reactive gas mixture. About 80% of imitation diamonds on the market are made of **cubic zirconia (CZ)**, a compound that is the cubic form of zirconium oxide, with yttrium oxide as a stabilizing agent. Both are white, opaque ores before being melted together by extremely high temperatures. CZ possesses the diamond's fire and brilliance, is nearly as hard but more brittle, and weighs 1.7 times as much. The brilliance of CZ can be dulled by contact with soap, make-up, and the natural oils of fingers. The best variety is colorless or white, but CZ is also produced in such colors as yellow, pink and black by adding chemical additives. It was discovered in 1937 by two German mineralogists, successfully grown in a laboratory in the 1970s by Soviet scientists, and first produced for mass consumption in the 1980s by an Austrian company.

Yttrium aluminum garnet (YAG) was introduced in 1969, and became a popular diamond imitation in the 1970s until overtaken by the newer CZ. Its popularity increased when the actor Richard Burton gave a 69-carat YAG stone to his wife, Elizabeth Taylor. YAG has been marketed under many names, including "Diamonaire" and "Diagem." It has less fire than a diamond and is heavier, but otherwise shows a good resemblance. It cuts and polishes like garnet.

▶ *Boules of synthetic ruby and synthetic sapphire.*

◀ *Large crystal of synthetic emerald.*

▼ *Yttrium aluminum garnet (YAG) gem.*

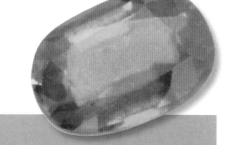

One of the most recent diamond imitations is called **moissanite**, which resembles diamond to such a degree that it can be falsely sold as the true gem. However, a trained gemologist using a microscope can readily spot the difference. Moissanite is doubly refractive, whereas diamond is singly refractive. It is made of laboratory materials with similar properties to silicon carbide, which was first discovered in 1893 by French scientist Dr Henri Moissan (1852–1907) in part of a meteorite in the huge Barringer Meteor Crater, Arizona. In 1995 the US company Cree developed a process for producing large, single crystals of silicon carbide. Moissanite is actually more brilliant than a diamond, has a higher luster, and is slightly less heavy.

Other diamond simulants have included colorless synthetic corundum, synthetic and natural spinel, synthetic and natural rutile, synthetic sapphire, synthetic beryl, quartz, zircon, topaz and glass (paste). Many other synthetic stones are on the market. For example, synthetic emerald is made with beryllium aluminum silicate, and synthetic alexandrite is produced using an oxide of beryllium and aluminum.

The imitations of organic gems include cultured pearls and pearls made by dipping alabaster glass beads into a solution of fish-scale essence and lacquer; tortoiseshell imitated by plastic; and coral simulated by staining vegetable ivory, bone, glass, porcelain, or plastic.

The manufacture of synthetic gems is big business. In 2006, the estimated production of synthetics in the US was US$47.4 million.

HISTORY OF SYNTHETICS

Imitation gems have been produced for nearly 7,000 years, with the ancient Egyptians using a colored ceramic to imitate turquoise. Laboratory gems, by contrast, were developed in the early 20th century, and today most are used as industrial gems. In 1902, the first commercially successful laboratory making synthetic gems was established by Auguste Verneuil (1856–1913). He made a synthetic ruby using powdered aluminum mixed with chromium oxide. Other creations have included synthetic spinel produced in 1908, synthetic sapphire created in 1911, and synthetic diamond made in 1955.

ORGANICS

▲ *Mother-of-pearl brooch with a cultured pearl at the center.*

Organic gems are formed by the biological processes of living organisms, and are therefore found on land and underwater. The main organic gems from animals are PEARL, IVORY, shell (including MOTHER-OF-PEARL and TORTOISESHELL), and CORAL, while those derived from plants are AMBER and JET. The most desired are pearls, which are the second best-selling gem after diamonds. Organic gems were the earliest type of carved gems, because their softness allowed them to be cut with simple tools. Ancient burial sites have yielded many artifacts made of these types of gems.

Secondary organic gems and collectables include **tagua nuts** and fossils such as **shark teeth** and **dinosaur eggs**. A fossil such as ammonite can be cut into gems, and other small fossil pieces are used in necklaces and bracelets. Tagua nuts, known as "vegetable ivory," are dried seed pods of the tagua palm tree found in the rain forests of Ecuador. Its outer skin covers a very hard egg-sized white interior that resembles ivory and can be carved like it. Artisans since the 19th century have produced items like jewelry pendants, inlays, buttons, and small figurines such as animals.

Organic gems are soft and porous. They scratch easily and absorb cosmetics, perfume and chemicals like hairspray, processes that could eventually devalue them. For these reasons, the gems should not be worn during intense activity and should not be dipped in chemical cleaners, but rather wiped clean with a moist cloth. It is also unwise to store them for lengthy periods in a closed container because dry air may cause the gems to crack.

◀ *Single-strand necklace of 21 baroque cultured pearls, joined by a pavé diamond clasp.*

A pearl is the lustrous, hard, and smooth rounded mass created inside all bivalve mollusks, but the only pearls prized as gem quality are in mollusks that produce them with nacre. Nacre is ARAGONITE, a calcium carbonate similar to MOTHER-OF-PEARL. Most pearls come from seawater mollusks, generally the large "pearl oysters" of the genus *Pinctada*, although they are not true oysters. Other pearl-producing mollusks are giant clams and giant conches, and pearls are also found in freshwater clams and mussels.

Pearls are formed when foreign matter like a grain of sand or a small parasite invades the mollusk, which envelops it within a soft body-part such as the mantle (tissue between the shell and body) which secretes nacre. A sac surrounds the irritant, and pearl building takes place within it. More nacre builds up a spherical or near-spherical "cyst" pearl – a pear-shaped pearl often taking about seven years. Sometimes an irritant is lodged in the mollusk's muscles, and the resultant pressure produces odd-shaped "blister" pearls.

Pearls are normally white or bluish-gray but can be pink or black, or dyed a range of other colors. The most highly valued are oriental pearls because of their iridescence, translucence, shape, and deep pearly luster created by diffraction and interference. Pearls are normally worn as earrings or necklaces, because they are comparatively soft and can be scratched with a coin, pin, or knife. They are also easily damaged by acids and heat.

Cultured pearls were first created by the Chinese in the 12th or 13th century, and the Japanese developed the industry in the late 1890s. These pearls are created by placing a small bead, usually made from mother-of-pearl shell, into the mantle of a three-year-old oyster for about four years. The nacre is laid down in parallel layers, while a genuine pearl has a concentric construction. If the nacre is too thin it will rub away.

The pearl was greatly valued in ancient China and India. The Greeks and Etruscans produced jewelry featuring pearls. The Romans sought them as prized gems and only citizens of a high rank were allowed to wear them. During the Renaissance, women wore pearls as necklaces, pendants, and earrings that dangled from gold wires. The largest known pearl is the Pearl of Laotze (or Pearl of Allah), weighing 14 pounds and 1 ounce (6.37 kilograms), found inside a giant clam in the Philippines in 1934.

Bead

◄ *Black and white cultured pearl ear pendant with marquise and circular-cut diamonds.*

▶ *Art Nouveau pendant, designed (c. 1900) by Georges Fouquet, featuring a pearl, opal and rose-cut diamonds.*

PEARL

Seawater pearls are found in subtropical and tropical waters, such as the Persian Gulf, especially from Oman to Qatar; the Gulf of Manaar in the Indian Ocean, which has the highly valued rose-colored variety; the coastal waters of East Africa from the island of Zanzibar to Mozambique; the Red Sea, although production has dwindled; the northwest of the Australian coast which has silvery white pearls; the islands of the South Pacific; and the US coasts of California and Florida. Freshwater pearls occur in the rivers of Europe and the United States, such as the Mississippi. "Scotch pearls" are in Scottish rivers like the Tay and Spey, and other pearls come from the River Conway in northern Wales.

▶ *Opaque oval pearl from China, 3.10 carats.*

Cabochon

Cameo

Polished

► *Pair of shell and gold earclips, designed as turbo shells by Seaman Scheoos and enhanced by sapphires.*

▲ *Ammolite consists of brilliant, intensely hued bits of the shell of ammonite, a fossilized mollusk. It is found only in southern Alberta, Canada. In appearance, it resembles black opal.*

Shells are composed of calcium carbonate and grow in all oceans. They are carved into cameos, used for inlay work, and made into buttons, ornaments, and jewelry like beads and necklaces. Shells with a pearly luster are ground back to the colorful, iridescent, nacreous interior layers and then they are polished.

Mother-of-pearl is nacre, the smooth calcium carbonate lining of iridescent luster found in some shells, such as the large pearl oysters, abalones existing in American waters, pearl mussels, and paua shells from New Zealand. Nacre is the substance of which pearls are made. The mineral ARAGONITE gives the luster to mother-of-pearl, while the color depends on the mollusk type and its location. Whiteness is produced by bleaching. Valued for its delicate beauty, a layer of mother-of-pearl is used for jewelry (often for children), inlay work, and decorative ornaments, as well as practical pieces, such as buttons and knife handles.

Shells having layers of different colors are also valued for cameos, with a figure in one color standing out against a background of another color. The layered giant **conch shell**, for example, is carved as cameos to reveal the two colors of pink and white. **Helmet shells** from the waters of the Caribbean are also used to create cameos. All of these low-relief carvings are curved, following the shape of the shell.

The **operculum** is also often used for jewelry. This material covers the shell opening of the large sea-snail from the tropical waters around Australia, Tahiti, Fiji, and Samoa. The operculum, from 0.2 to 3 inches (5 to 75 millimeters), is dome shaped with circular markings of green, brown, and white. Underneath, it has a spiral pattern of growth. Another popular shell is the larger **top shell**. It is a low conical shape and has a pearly interior. They are sometimes made into beads and strung as necklaces.

▲ *Mother-of-pearl shell from China, 45.10 carats.*

SHELL (MOTHER-OF-PEARL)

The iridescent color of the mother-of-pearl of the abalone shell in the Caribbean is predominantly red and pink. Abalone is also called ear-shells *ormer* in the Channel Islands of Jersey and Guernsey between England and France, *perlemoen* in South Africa, and *paua* in New Zealand. The iridescence of the paua shell is blue and green.

Tortoiseshell is not the shell of a tortoise; in fact it is the carapace (shield) of the hawksbill sea turtle (*Eretmochelys inbricata*). The hawksbill is a tropical marine turtle, found around coral reefs and lagoons in the Atlantic, Pacific, and Indian oceans. Hawksbills nest throughout the tropics, with important rookeries at sites in Queensland in Australia, the Seychelles, Caribbean, and Mexico. The carapace resembles horn, but is harder and more brittle. The shells must be flattened and can be pressed together by a heat process. It is then polished and cut.

The marbled pattern of several colors on a yellow background and the deep translucence of the shield have long been used to create jewelry and decorative items like combs, frames for glasses, necklaces, brooches, clocks, ashtrays, lamps, and knife handles. Tortoiseshell has also been valued since Roman times as an inlay for furniture, although imitative plastic is generally used today. The Romans were introduced to the shells by the Egyptians. In the 17th century, the French turned tortoiseshell carving into an art, adding designs in gold, silver, and mother-of-pearl.

◀ *The marbled pattern of polished and cut tortoiseshell. The sale of the sea turtle's shell is banned.*

SHELL (TORTOISESHELL)

Hawksbill turtles have been an endangered species since 1973, and an international treaty bans the sale of the shells. This has nearly killed the thousand-year-old tradition in Japan of carving tortoiseshell (*bekko* in Japanese). Before the ban, the Japanese annually imported about 33 tons of hawksbill shell, which is equivalent to 31,000 turtles.

▲ *The hawksbill turtle is endangered for two reasons – its eggs and flesh are regarded as a delicacy, and its carapace has been prized since ancient times as the source for tortoiseshell.*

▶ *Tortoiseshell, diamond, and ruby brooch designed by Nardi. The carved tortoiseshell hand is modeled on the hand of the Dogaressa, wife of the Doge (historic ruler of Venice).*

Coral is made by small and simple marine invertebrate animals such as the coral polyp, which is related to sea anemones. Their dense limestone (calcium carbonate) skeletons cluster together and build up over thousands of years into coral reefs and coral islands. Corals can exist as single, usually cone-shaped coralites, commonly called horn corals. Most corals, however, are joined together in massive colonies. The most abundant fossil corals are scleractinian corals in reefs. As more animals are created by budding, lower individuals die and their skeletons add new layers that can form fringing reefs along the shore, barrier reefs offshore, and atolls (circular reefs enclosing a lagoon). Marine plants like algae also add their skeletal structures to the reefs.

Coral reefs support living corals and many varieties of plant and animal life. The most famous reef is the Great Barrier Reef off northeastern Australia, which extends for more than 1,250 miles (2,000 kilometers). The reefs are now facing environmental damage from global warming, water pollution, coastal development, overfishing, and tourism. In 1998 the El Niño climate event destroyed about 16% of the world's reefs in nine months. More than 60 countries suffered coral bleaching, in which high temperatures expel the algae from the reefs. The most damaged reefs are in the Persian Gulf, Indian Ocean, Southeast and East Asia, and the Atlantic Ocean and Caribbean. The healthiest reefs are in the Pacific Ocean and off Australia. Research by the Global Coral Reef Monitoring Network predicts that 60% of the world's reefs could be lost by 2030 if warming continues. Coral harvesting has become more strictly regulated and coral more expensive.

Precious coral has long been used for jewelry and decoration. The Romans believed **red Mediterranean coral** had magical and medicinal properties, and Roman children wore coral necklaces as protection from dangers. The Gauls used coral to

▲ *Coral, pearl and gold cameo of Bacchus, designed (c. 1854) by François Désiré Froment-Meurice.*

CORAL

Most coral types exist in warm waters from 20 to 1,000 feet (5 to 300 meters) deep, with the best-quality coral found from 100 to 160 feet (30 to 50 meters). Red and pink corals exist along the Mediterranean coasts of Italy, France, and Africa. Black and golden coral occur off Hawaii, Australia, and the West Indies. Shining black coral comes from the Indian Ocean, while pink, red, and white coral comes from Japan. Other deposits are found in the Red Sea, and off the coasts of Algeria, Tunisia, and Malaysia.

▶ *Coral branches in their natural state and polished examples.*

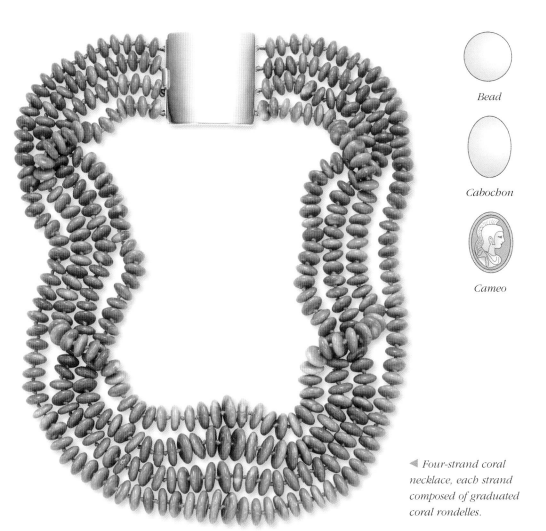

Bead

Cabochon

Cameo

◄ Four-strand coral
necklace, each strand
composed of graduated
coral rondelles.

decorate their helmets and weapons. Victorian babies from wealthy families had teething rings of coral. Native American artisans, like the Navajo, Hopi, Zuni, and Pueblo, also created fine coral pieces.

Today, craftsmen, especially in Italy, have developed great skills in cutting, carving, and polishing coral. Precious red and white corals are especially used for jewelry. Pieces are cut as cabochons and many are faceted and fashioned as beads. Coral with different shapes is perforated to be strung as necklaces and bracelets. Domed pieces are formed to set in brooches, and others are used in earrings, pendants, pins, and rings, as well as being carved into small figurines and cameos.

Black coral is also known as *Akabar* or "King's coral," and **blue coral** is also called *Akori*. Red, pink, white, and blue corals are made of calcium carbonate, while black and golden corals are formed from a material known as conchiolin. All corals have a delicate graining of stripes or spots in their skeletal structure.

▲ Polished cabochon
of coral from the
Philippines, 20.25
carats.

231

Ivory is the hard white dentine from the tusks of elephants and some other animals, such as walruses, sperm whales, and narwhals. Only the elephant's ivory is large enough to be of good commercial value. This thickened dental enamel has a rich creamy color, excellent texture, and is elastic. Ivory from African elephants has a transparent mellow color with almost no grain or mottling. Indian elephants, which have smaller tusks (or none at all), produce ivory that is more white but also yellows more easily. Most ivory is soft enough to be carved with ordinary wood-carving chisels, but hard ivory is found in the western half of Africa. Ivory is very durable, is difficult to damage or destroy, does not burn, and resists water. Artisans try to retain the original shape of the ivory piece if possible.

Due to CITES restrictions (*see panel below*), several materials have been used to replace ivory. These include plastic, bone, horn, and vegetable ivory, which is the hard albumen of the seeds of palm trees such as the ivory palm from Peru and the doum palm of the Nile region of Africa.

Carved ivory has been greatly prized for millennia. Carvings estimated to be about 30,000 years old have been found on mammoth ivory in French caves. The Chinese carved fine ivory figures in the Han dynasty (206 BC–AD 220). The ancient civilizations of Egypt, Babylon, Japan, India, Greece, and Rome produced works in ivory. Even the plague in Britain during the Roman occupation may have been caused by rats on ships carrying the popular ivory to the Romans. Between the 7th and 12th centuries, a very sophisticated tradition of ivory carving developed in Britain, and by the 13th century

▲ *Art Nouveau ivory, amethyst and enamel pendant and necklace, designed (c. 1905) by René Lalique.*

IVORY

Because of ivory's great popularity, elephants have been slaughtered for their tusks to the extent that their very existence in Africa is threatened. The African elephant is the major source of ivory, since male Asian elephants have smaller tusks and Asian females are tuskless. The African elephant population fell from 1.3 million in 1979 to 609,000 in 1989. This led to an international treaty in 1989 banning the ivory trade. The treaty was established by the Convention on International Trade in Endangered Species (CITES), a United Nations agency. Before the 1989 CITES ban, illegal and legal ivory exports amounted to 770 tons, or 75,000 elephants. In 1997, CITES partially eased the restrictions to allow Botswana, Namibia, and Zimbabwe to cut legal stockpiles. In 2002, it permitted Namibia, South Africa, and Botswana to sell up to 30 tons of their stockpiles. Illegal poaching still continues today. China and Japan are the leading export markets for ivory.

Bead

Cameo

Polished

◀ *Two men, amid large piles of elephant tusks, measure and weigh a tusk. The ivory was obtained from an elephant cull in Kruger National Park, South Africa.*

▼ *Ivory caskets from the late 19th century.*

Europeans began to create highly elaborate ornamental carvings. Ivory has also long been used for inlay work and as a fine surface for miniature paintings.

The modern uses of ivory have included piano keys, fine-toothed combs, chess pieces, pistol handles, and figurines. Billiard balls were traditionally made of ivory, but synthetic materials are now used for these and other items. The Japanese regard ivory as a precious material and have employed it for ornamental buttons and other decorative and functional items. Ivory is also a good heat insulator, and is used in teapots, coffeepots, and some electronic equipment.

Ivory from other animals has had periods of popularity. Walrus tusks were often carved in the Middle Ages in northern European lands, and hippopotamus teeth were widely used in the 18th and 19th centuries in England and France. Eskimos in North America carved ivory from walrus tusks, which were also used along with whale teeth by sailors for their carvings ("scrimshaws").

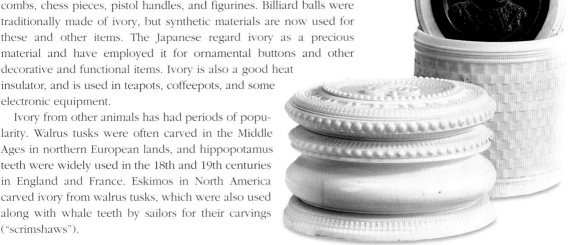

Cabochon

Cameo

Polished

Warm, glowing and imbued with the traces of millions of years in Earth's history, amber has attracted attention since prehistoric times. Stone Age amber artifacts are on display in museums around the world. In classical times, it was used medicinally and was also believed to offer a magical light for the deceased as they progressed through the underworld. From the 14th century, guilds of craftsmen specialized in creating jewelry and ornaments, both large and small, from amber.

Today, it is sought after by collectors and jewelry lovers for its appearance, as well as by scientists pursuing knowledge of the ancient history of the Earth.

Amber is not a mineral. Rather it is organic material, the fossilized resin of ancient trees formed through natural polymerization of the original organic compounds. It is usually found within Cretaceous or Tertiary sedimentary rocks, especially clays, shale and sandstones and associated with lignite (a woody, brown coal formed from the remains of the tree that produced the resin). Amber is also found washed up on the seashore, because its relative density is such that it floats in salt water.

Rubbing amber with a cloth will generate static electricity, causing paper to stick to it. This property has been recognized since ancient times, and in fact the word "electricity" derives from the ancient Greek word for amber, *elektron*, meaning "sun-made."

Amber is very prone to inclusions, because the resin from which it is formed is a sticky substance, produced by trees as a defense against wood-boring insects. Resin entraps not only insects, like flies and bees, but also larger living things, from grasshoppers to lizards and even small frogs. More than 1,000 extinct species of insect have been identified from remains found in amber. Amber also trapped flowers, mushrooms and leaves, as well as hair and feathers. These types of inclusions greatly increase the value of an amber specimen, and have more bearing on price than color or clarity. Liquid and air bubbles can also be included, but when amber is to be used in jewelry, these are often removed by boiling it in rapeseed oil.

AMBER	
CHEMICAL COMPOSITION	78% CARBON, 10% HYDROGEN, 11% OXYGEN
COLOR	YELLOW, ORANGE, RED, WHITE, GREEN, BROWN, BLUE
REFRACTIVE INDEX	1.54
RELATIVE DENSITY	1.05–1.10
HARDNESS	2–3
CRYSTAL GROUP	AMORPHOUS
CLEAVAGE	NONE
FRACTURE	CONCHOIDAL
TENACITY	TOUGH TO BRITTLE
LUSTER	RESINOUS, GREASY
TRANSPARENCY	TRANSPARENT TO OPAQUE
DISPERSION	NONE
BIREFRINGENCE	NONE
PLEOCHROISM	NONE

LOCATION Burma (Myanmar), Dominican Republic, Estonia, Italy, Latvia, Lithuania, Netherlands, Romania, Russia, United Kingdom, United States.

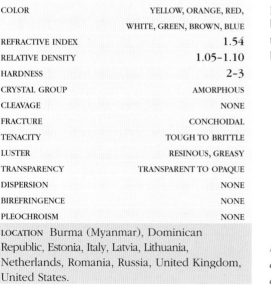

▲ *Various facets and colors of amber.*

AMBER

Until the mid 19th century, most amber, then often known as "seastone," was picked up along the seashore. From the mid 19th century, dredging operations began along the shores of the Baltic, and mining, often open-pit, started in the late 19th century. Today, the Baltic states (Estonia, Latvia, and Lithuania) remain an important source of amber. The Curonian Spit in western Lithuania is known as the "Amber Coast." The Dominican Republic has amber mines close to its capital, Santo Domingo, and the city of Santiago.

▲ *Insect in fossilized resin (amber).*

Amber is found in a range of colors. The most common, and most popular, falls into the yellow to reddish-orange range, but amber can also be white, green or even blue. Some scientists believe that color is in a large part due to the source tree: for example, pines produce golden-yellow amber, while deciduous trees are responsible for the more reddish material. A green color is due to residues of other decomposed organic materials. Most amber will darken to a mellow brown after long exposure to air.

There are two common classification schemes for amber. One is based on the source location. For example, Romanian amber is **rumanite**, Sicilian amber is **simetite**, and Burmese amber is **burmite**. Baltic amber is known as **succinite** (because it contains succinic acid). The other scheme is based on appearance. Amber is judged on its transparency, from clear to cloudy. Cloudy amber is then subdivided into fatty amber (translucent yellow, with inclusions of suspended dust particles); bony or osseous amber (opaque and whitish to brown); and foamy (opaque, very soft, often with inclusions of pyrite).

Amber requires great care. It is extremely heat-sensitive – it has been burned as incense for centuries – and must be protected from heat sources, including hot water and even strong sunlight, which can cause it to dehydrate. It is also very sensitive to hairspray and perfume, and contact can produce white encrustations. Because it is soft, it should also be protected from bumps and scratching: store separately from other jewelry, and string beads of amber with knots in between. Ultrasonic or steam cleaning can cause amber to shatter – it should be cleaned with warm (not hot) water and a soft cloth.

There is much imitation amber on the market, made of glass or plastic or even modern tree resins. Amber feels warm to the touch and will float in salt water. The residues from cutting, as well as small pieces too small to cut or polish, are heat processed to create **ambroid**, or pressed amber. In such pieces, the translucence tends to be variable, and internal air bubbles are sometimes elongated due to the high pressure. Ambroid may also be artificially colored.

▲ *The different colors of amber may be due to different source trees.*

Bead

Cameo

Polished

▼ *Slab, cuts and ring made of jet.*

Jet, like diamond, is made primarily of carbon. But there the similarities end, because jet is black and opaque, while diamond is clear and glittering.

Jet is usually found in carbon-rich beds in shale strata. It occurs in tubers rather than veins, and can also be found washed up on the seashore. Jet is a type of brown coal, a fossilized wood of an ancient species of *Araucaria*, a conifer tree similar to the present-day monkey-puzzle tree. These trees flourished in the Jurassic period about 180 million years ago. When the trees died, some fell into swamps or rivers, broke up, and were eventually carried out to sea. As the trunks and branches became waterlogged they sank to the seabed. Great pressure from layers of organisms, mud and detritus flattened the tree fragments and, together with chemical changes, altered the wood to jet. Analysis of the oil in hard jet confirms that it was formed under seawater, while it is probable that soft jet was formed under freshwater. Jet's wood origins are sometimes manifested in a grain-like texture, although most jet is smooth.

Used since Palaeolithic times, jet has had numerous applications. Many cultures believed it offered medicinal benefits, especially when the smoke released by burning was inhaled. Polished jet produces such a shine that in medieval times it was used as a mirror. Although set into jewelry since the ancients, it was not until Queen Victoria wore jet after the death (1861) of Prince Albert that it became the standard for mourning jewelry, particularly in England.

Jet is commonly faceted for jewelry, and it can take a high polish. Also, because it is quite light, it is comfortable to wear. However, care must be taken that it does not dry out, because dehydration causes the surface to crack.

Jet has many imitators, both natural, such as obsidian, and some artificial, like plastic or glass. Real jet always feels warm to the touch. It can also become electrically charged if rubbed with a cloth (although this is also true of some plastics).

JET

CHEMICAL COMPOSITION	60–75% CARBON
COLOR	BLACK
REFRACTIVE INDEX	1.66
RELATIVE DENSITY	1.30–1.35
HARDNESS	2–4
CRYSTAL GROUP	AMORPHOUS
CLEAVAGE	NONE
FRACTURE	CONCHOIDAL
TENACITY	TOUGH TO BRITTLE
LUSTER	VELVET-LIKE
TRANSPARENCY	OPAQUE
DISPERSION	NONE
BIREFRINGENCE	NONE
PLEOCHROISM	NONE

LOCATION England, France, Germany, Portugal, Russia, Spain, Turkey, United States.

JET
Today, jet is rarely used in jewelry. Depending on the source, it can be carved into ornaments -- the jet from Whitby, Yorkshire, north England, is generally acclaimed to be of the best quality for carving and jewelry.

PRECIOUS METALS

The main precious metals of gold, silver, and platinum have long been valued for their beauty, rarity, and as a protection against international economic collapse. Throughout history, alchemists have attempted to create gold and silver by transforming base metals, such as lead and copper. For millennia, gold and silver have been used as mediums of exchange, and this has more recently included the rarer platinum. All of the platinum that has been mined could fit into a 20-foot (6-meter) cube. Other members of the platinum group of silver-white metals are palladium, rhodium, iridium, ruthenium, and osmium.

Precious metals are used for expensive jewelry. Two-tone jewelry is often created, such as combining yellow gold with white platinum for rings. Platinum is used to hold diamonds and other precious gems because it is the heaviest of the precious metals, weighing nearly twice as much as gold. All these metals can be scratched, so the jewelry should not be worn while doing vigorous work.

Gold, silver, and platinum can be bought as bullion bars, but investors prefer coins. Besides the value of precious metals for jewelry, coins, and medals, they have many industrial uses, as in the production of resistors and objects that must resist corrosion, such as electrodes. Other uses include dental work (gold), thermocoupling wire (platinum), and catalytic converters to reduce car exhaust emissions (platinum, palladium, and rhodium).

Precious metals are measured in troy weight, which has units of pennyweight, ounces, and pounds, although the latter two are not equivalent to the more commonly used measures of the same name. Gold is also measured in metric grams.

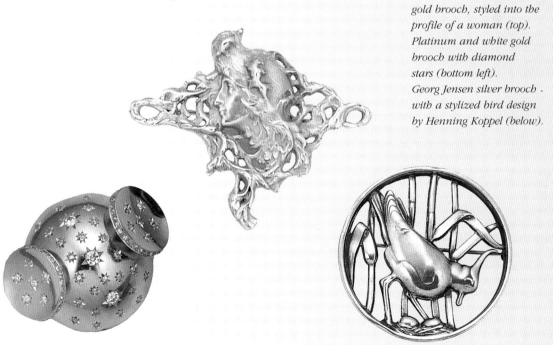

◄ *French Art Nouveau gold brooch, styled into the profile of a woman (top). Platinum and white gold brooch with diamond stars (bottom left). Georg Jensen silver brooch with a stylized bird design by Henning Koppel (below).*

Gold was used more than 5,000 years ago in Mesopotamia, and has never lost its attraction. A symbol of divinity, used for religious objects; of royalty, used for the crowns and scepters of kings and queens; and an indication of wealth, its gleam adorning the necks of the richest men and women throughout history – gold has been valued for thousands of years.

The esteem in which gold has always been held is demonstrated by the fact that it has been accepted in exchange for goods and services by all cultures throughout history. In fact, up until World War I, the gold standard was the basis for the world's currencies, and even today, 45% of the world's gold is held in government reserves.

The element gold is found in low concentrations in igneous rocks, and it rarely combines with other elements. It is often found in association with copper and lead, and because it does not combine, it is easily recovered. It also occurs in small amounts in hydrothermal veins, in association with quartz and pyrite. It is found in alluvial deposits, in which gold, because of its density, is separated from other minerals during weathering and transported to become concentrated in stream and other sediments, which may be loose and unconsolidated or hardened into rock. Tiny grains of gold can be carried long distances by streams and can be recovered from gravel by panning, which involves washing away all but the heavy minerals and then searching for any flecks of gold.

Because gold is so soft, it is alloyed with other metals (usually silver, copper and zinc) to increase its hardness so it can be used in jewelry. White gold has a higher proportion of silver than yellow gold does. Gold purity is defined by its carat value – one carat is equivalent to 1/24th, with 24-carat gold being pure gold. Most jewelry is either nine-carat or 14-carat gold.

▲ *Gold and gem set brooch, designed (c. 1940) as a butterfly with wings studded with diamonds and synthetic rubies and sapphires.*

▼ *Gold nugget.*

▶ *Gold on quartz.*

GOLD

The market for gold in Europe exploded with the Spanish conquest of the New World in the 15th century, with the height of gold production in the 1490s. The New World mines continued to be productive through the early 1800s, when they were overtaken by Russia as the leading source of gold. Since the end of the 19th century, however, South Africa has dominated the market and is today the source of around one-third of total global production. Russia, the United States of America, and Australia together account for a further one-third.

◄ The Sovereign's Orb in the British crown jewels was made for Charles II's coronation in 1660. The hollow orb is made from unmarked gold and set with more than 600 precious stones and pearls. It weighs 42 ounces (1.32 kilograms) and is 6.5 inches (16.5 centimeters) in diameter.

▲ Cartier ring with three gold bands – one set with diamonds, one set with rubies, and one set with sapphires.

Gold is remarkable for its ductility and malleability – in fact, it is the most malleable metal of all. For example, 0.6 ounces (20 grams) of gold can be stretched into a wire more than 0.6 miles (1 kilometer) long or beaten into a sheet of gold leaf more than 210 square feet (20 square meters) in area. In addition to this, it does not tarnish or corrode because it is chemically inert.

Its durability and workability have inspired crafts-people and their patrons over the millennia. Beautiful jewelry and ornaments were created by the earliest cultures, and museums around the world display gold objects created by Egyptians, Minoans, Assyrians, and Etruscans. During the Middle Ages, gold objects and gold-leaf veneers adorned elaborate churches, and during the Renaissance, gold jewelry was worn by the most important people. Today, although gold is no longer the province of only the wealthy, because it is more affordable, it is no less precious for it.

GOLD	
CHEMICAL COMPOSITION	AU
COLOR	GOLD-YELLOW
REFRACTIVE INDEX	NONE
RELATIVE DENSITY	19.3
HARDNESS	2.5
CRYSTAL GROUP	CUBIC
CLEAVAGE	NONE
FRACTURE	HACKLY
TENACITY	TOUGH YET MALLEABLE
LUSTER	METALLIC
TRANSPARENCY	OPAQUE
DISPERSION	NONE
BIREFRINGENCE	NONE
PLEOCHROISM	NONE

LOCATION Worldwide, especially Australia, Brazil, Canada, China, Indonesia, Russia, South Africa, United States.

Silver has been widely used for jewelry and ornaments, as well as for money. Its white gleam enhances the appearance of transparent gems, encouraging the reflection of light, and its softness means that it can be worked in great detail.

Silver is widely distributed. However, compared to other metals, it is relatively scarce. It is present in many minerals and occurs in hydrothermal veins, or in small amounts in the oxidized zone of silver-bearing ore deposits. More than 80% of silver produced is recovered as a by-product of mined copper, lead, or zinc. Native silver normally occurs as nuggets or grains, although fine dendritic (branch-like) forms are also sometimes found.

▶ *Arts and Craft silver brooch with cabochon-cut moonstone and sapphire gems and red paste.*

▲ *Silver found in pure form in the ground is called native silver – its structure is dendritic (branch-like).*

In the periodic table, the element silver is located between gold and copper, and its properties are intermediate. It ranks second only to gold in terms of malleability and ductility, making it easy to work. It has the highest electrical and thermal conductivity of any metal, so it is also widely used industrially, especially for electronic circuits and conductors. Silver contacts inside switches turn on and off the powerful electric current that flows into our homes, our lamps and our appliances. Silver halides (salts) are photosensitive and the photographic industry is the largest single end-user of silver. Approximately 2,000 color photographs can be taken using 0.3 ounces (10 grams) of silver. Silver's properties make it ideal for use as a catalyst in oxidation reactions; for example, the production of formaldehyde from methanol and air by means of silver screens or crystallites contains a minimum 99.95 weight-percent silver. Silver is also used to make mirrors.

The discovery that many liquids stay pure longer in silver vessels led to its desirability as a container for long voyages. The Greek historian Herodotus (*c.* 485 BC–*c.* 425 BC) noted that Cyrus the Great, King of Persia (550–529 BC), had water drawn from a special stream, "boiled, and very many four-wheeled wagons drawn by mules carry it in silver vessels, following the king wheresoever he goes at any time."

The phrase "born with a silver spoon in his mouth" refers not to wealth but to health. In the early 18th century, babies fed with silver spoons were found to be healthier than those fed with spoons made from other metals, and silver babies' dummies became widely used in the United States because of their beneficial health effects. Silver also has a variety of uses in pharmaceuticals. For instance, silver sulfadiazine is the most powerful compound for burn treatment. Silver is also used as a bactericide in many water purification systems

The first silver artifacts include ornaments and jewelry found in Sumerian royal tombs, dating back to 4000 BC. By 2000 BC, mining and smelting of silver-bearing lead ores was already common in

SILVER	
CHEMICAL COMPOSITION	AG
COLOR	SILVER-WHITE
REFRACTIVE INDEX	NONE
RELATIVE DENSITY	10–12.0
HARDNESS	2.5
CRYSTAL GROUP	CUBIC
CLEAVAGE	NONE
FRACTURE	HACKLY
TENACITY	TOUGH YET MALLEABLE
LUSTER	METALLIC
TRANSPARENCY	OPAQUE
DISPERSION	NONE
BIREFRINGENCE	NONE
PLEOCHROISM	NONE

LOCATION Australia, Canada, Chile, Mexico, Peru, Russia, United States.

Europe and the Middle East. But although it is, after gold, the most widely used metal in jewelry today, silver was rarely used in jewelry in pre-Classical times. Instead, it was an important material for ornamental metalwork; it was never used for weaponry because it was too soft. Roman silverwork is especially noteworthy.

Silver has also long been used for money, and coinage made of silver has been found in the Nile region of Africa and the Indus River valley of Asia dating back to 800 BC. In AD 1995, the United States Treasury issued more than 6.7 million troy ounces of silver coins.

Because native silver is very soft, it has to be alloyed with another metal, usually copper, to make it tough enough for use in jewelry, coinage, and ornaments. The purity of silver is defined as its "fineness," in parts per thousand. For example, **sterling silver** is 925 fine. In many countries, sterling silver (92.5% silver, 7.5% copper) is the standard for silverware and has been since the 14th century. Jewelry-grade silver is 800 fine. Silver itself is added to gold as an alloy, to harden it and allow it to be used for jewelry – jewelry-grade gold can contain around 25% silver.

When freshly polished, silver has a distinctive gleam and silver-white color. However, on exposure to oxygen, silver forms silver oxide, a black layer that tarnishes the surface. This is easily removed with soap and water or a special silver cleaner. However, some tarnish can serve as a contrast to the natural brightness of polished silver, enhancing the detail of fine metalwork.

◀ *Silver necklace with rutilated quartz pendant.*

▼ *Intergrown silver and copper.*

SILVER

Silver mining began early in the New World, particularly in South America. Following the Spanish discovery of the New World, Europeans quickly began to exploit the silver mines in Mexico, Bolivia, and Peru, answering a demand for silver in Europe inspired by the fine craftsmanship of the Renaissance. The first major exploitation of New World silver was in the Potosí district of southern Bolivia.

The South American mines dominated the world market until the mid 19th century, when large deposits of silver were found in the US, most notably the Comstock Lode area in Nevada, the Leadville district in Colorado, and various districts in Utah. (Nevada is sometimes known as the "Silver State.") By the end of the 19th century, high-grade ores throughout the world had been largely depleted, but technological improvements allowed for increased production, and by 1920 about 190 million troy ounces of silver were being produced annually. Today, Mexico and Peru have overtaken the United States to become the dominant silver producers.

▼ *Ancient silver coin showing portrait of Philip II of Macedon (382–336 BC).*

241

In 1790, King Louis XVI of France declared platinum to be "the only metal fit for kings." Although it is a relatively recent scientific discovery, platinum has quickly established itself as the prestige metal for precious jewelry. It also has many industrial and medical uses.

Platinum is a chemical element. While it has been known since ancient times, it was only isolated scientifically in 1735, because, unlike gold, for example, platinum requires complex chemical processing for isolation and identification. The grains of platinum are usually too small to be seen without magnification. It is not the major metal in any ores, but it is usually disseminated in sulfide ores in igneous rocks. There are also a few small alluvial deposits, in which platinum generally occurs as grains rather than nuggets. Mining and processing is labor-intensive – it takes 10 tons of ore and five months to produce 1 ounce (30 grams) of pure platinum. Because platinum has a high melting point, it is a challenge to work – in fact, it was not until the 1920s that technology was developed to work the metal conveniently. However, this means that it can be used in a quite pure state. The platinum standard is measured at 1,000 parts equaling 100%. Most platinum jewelry is rated at 95% (950) or 90% (900) pure, with the balance usually consisting of copper.

Platinum has been used for jewelry and ornamentation for thousands of years. The Egyptians decorated sarcophagi with platinum fittings, and in South America the Spanish *conquistadors* found platinum as they were panning for gold. Believing it to be unripe silver, they named it *platina*, meaning "little silver," and threw it back in the river.

Only around 40% of platinum is used in jewelry. About the same amount is used in catalytic converters, and the rest is used for industrial and medical applications such as computer hard disk drives, LCD displays, and anti-cancer drugs.

Platinum's bright white luster and good weight – it is almost twice as heavy as gold – suit it as a setting for precious gems. Platinum resists tarnish and is extremely hard and hypoallergenic. Over the years, any scratches to platinum are absorbed into an attractive patina that replaces a high polish.

▼ *Platinum nugget.*

▲ *Platinum and diamond wristwatch, Swiss, c. 1925.*

PLATINUM

Malleable platinum was not produced until 1789, and soon jewelry made of the rare and precious metal was much in demand for European monarchs. In 1824 large deposits were discovered in the Ural Mountains of Russia. Platinum prices peaked in 1912, when white gold was invented as a cheaper substitute. In 1924, Dr Hans Merensky (1871–1952) discovered the world's largest deposits of platinum at the Bushveld Complex, Transvaal, South Africa. In 1939, the US government declared platinum a strategic metal and banned its use in jewelry for the remainder of World War II.

PLATINUM

CHEMICAL COMPOSITION	PT
COLOR	GRAYISH WHITE
REFRACTIVE INDEX	NONE
RELATIVE DENSITY	21.45
HARDNESS	4
CRYSTAL GROUP	CUBIC
CLEAVAGE	——
FRACTURE	CONCHOIDAL
TENACITY	TOUGH
LUSTER	METALLIC
TRANSPARENCY	OPAQUE
DISPERSION	NONE
BIREFRINGENCE	NONE
PLEOCHROISM	NONE

LOCATION Australia, Borneo, Brazil, Canada, Colombia, Ecuador, Finland, Ireland, Madagascar, New Zealand, Peru, Russia, South Africa, United States.

IMPORTANT GEM LOCATIONS OF THE WORLD

MAP CONTENTS

On the following pages we show the locations of 30 of the more important gems from around the world.

GEMS OF THE WORLD – QUICK REFERENCE CHART

agate

alexandrite

amber

amethyst

aquamarine

chrysoberyl

citrine

coral

diamond

emerald

garnet

heliodor

jadeite

jasper

lapis lazuli

malachite

moonstone

nephrite

opal

pearl

peridot

rhodochrosite

ruby

sapphire

spinel

topaz

tourmaline

turquoise

zircon

zoisite

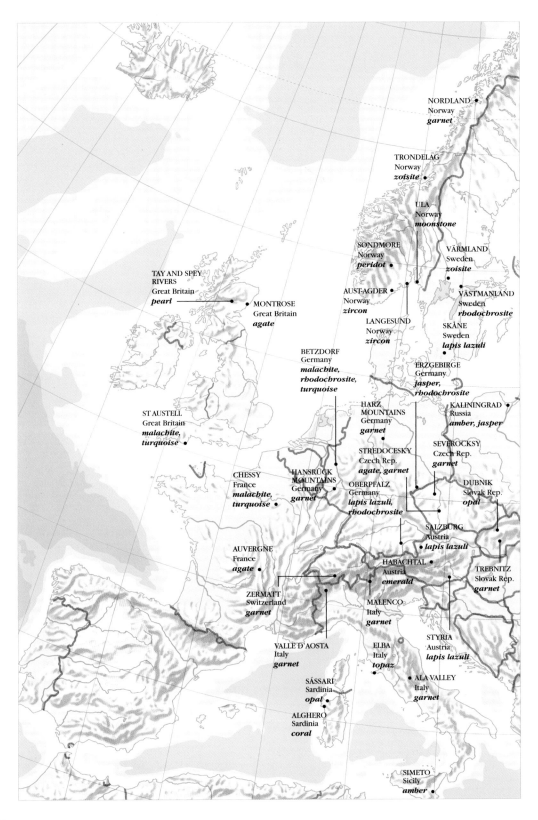

NORDLAND
Norway
garnet

TRONDELAG
Norway
zoisite

ULA
Norway
moonstone

VÄRMLAND
Sweden
zoisite

SONDMORE
Norway
peridot

TAY AND SPEY
RIVERS
Great Britain
pearl

MONTROSE
Great Britain
agate

AUST-AGDER
Norway
zircon

VÄSTMANLAND
Sweden
rhodochrosite

LANGESUND
Norway
zircon

SKÅNE
Sweden
lapis lazuli

BETZDORF
Germany
malachite,
rhodochrosite,
turquoise

ERZGEBIRGE
Germany
jasper,
rhodochrosite

KALININGRAD
Russia
amber, jasper

ST AUSTELL
Great Britain
malachite,
turquoise

HARZ
MOUNTAINS
Germany
garnet

STREDOCESKY
Czech Rep.
agate, garnet

SEVEROCKSY
Czech Rep.
garnet

CHESSY
France
malachite,
turquoise

HANSRÜCK
MOUNTAINS
Germany
garnet

OBERPFALZ
Germany
lapis lazuli,
rhodochrosite

DUBNIK
Slovak Rep.
opal

AUVERGNE
France
agate

SALZBURG
Austria
lapis lazuli

HABACHTAL
Austria
emerald

TREBNITZ
Slovak Rep.
garnet

ZERMATT
Switzerland
garnet

MALENCO
Italy
garnet

VALLE D'AOSTA
Italy
garnet

ELBA
Italy
topaz

STYRIA
Austria
lapis lazuli

SÁSSARI
Sardinia
opal

ALA VALLEY
Italy
garnet

ALGHERO
Sardinia
coral

SIMETO
Sicily
amber

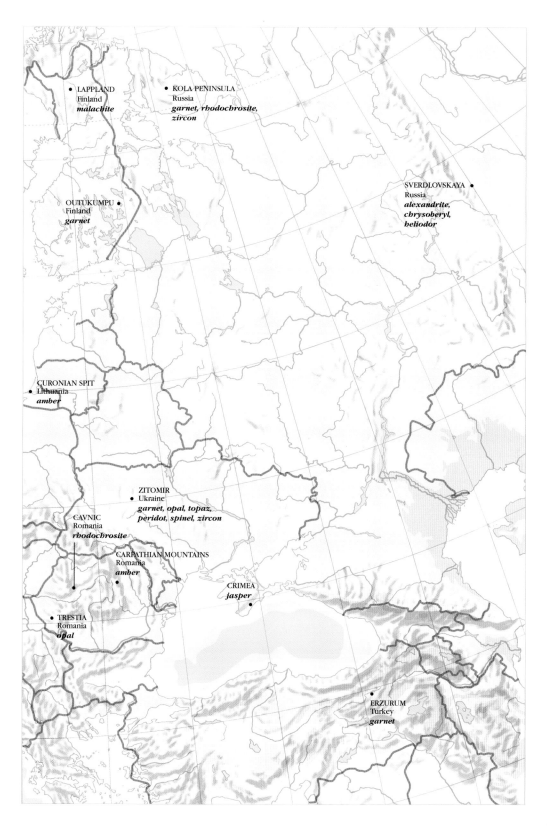

LAPPLAND
Finland
malachite

KOLA PENINSULA
Russia
*garnet, rhodochrosite,
zircon*

SVERDLOVSKAYA
Russia
*alexandrite,
chrysoberyl,
heliodor*

OUTUKUMPU
Finland
garnet

CURONIAN SPIT
Lithuania
amber

ZITOMIR
Ukraine
*garnet, opal, topaz,
peridot, spinel, zircon*

CAVNIC
Romania
rhodochrosite

CARPATHIAN MOUNTAINS
Romania
amber

CRIMEA
jasper

TRESTIA
Romania
opal

ERZURUM
Turkey
garnet

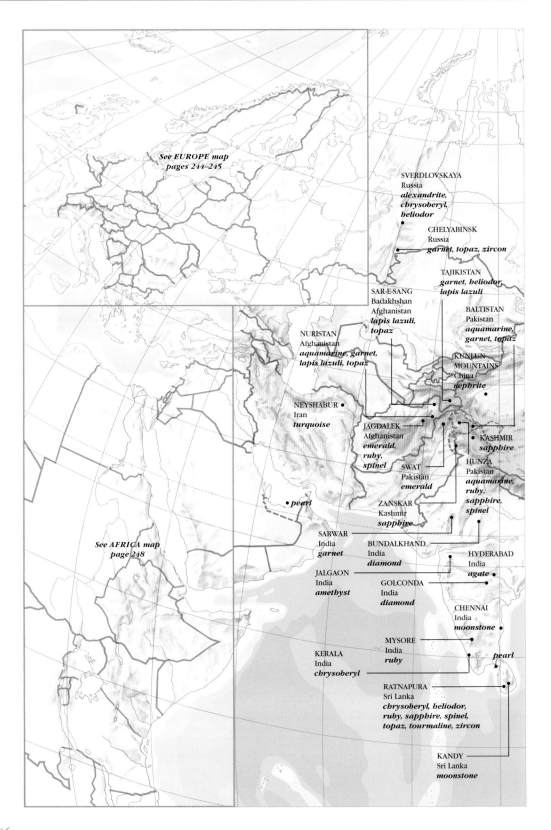

SVERDLOVSKAYA
Russia
*alexandrite,
chrysoberyl,
heliodor*

CHELYABINSK
Russia
garnet, topaz, zircon

TAJIKISTAN
*garnet, heliodor,
lapis lazuli*

SAR-E-SANG
Badakhshan
Afghanistan
*lapis lazuli,
topaz*

BALTISTAN
Pakistan
*aquamarine,
garnet, topaz*

NURISTAN
Afghanistan
*aquamarine, garnet,
lapis lazuli, topaz*

KUNLUN
MOUNTAINS
China
nephrite

NEYSHABUR •
Iran
turquoise

JAGDALEK
Afghanistan
*emerald,
ruby,
spinel*

KASHMIR
sapphire

HUNZA
Pakistan
*aquamarine,
ruby,
sapphire,
spinel*

SWAT
Pakistan
emerald

• *pearl*

ZANSKAR
Kashmir
sapphire

SARWAR
India
garnet

BUNDALKHAND
India
diamond

HYDERABAD
India
agate •

JALGAON
India
amethyst

GOLCONDA
India
diamond

CHENNAI
India
moonstone

MYSORE
India
ruby

KERALA
India
chrysoberyl

pearl

RATNAPURA
Sri Lanka
*chrysoberyl, heliodor,
ruby, sapphire, spinel,
topaz, tourmaline, zircon*

KANDY
Sri Lanka
moonstone

*See EUROPE map
pages 244–245*

*See AFRICA map
page 248*

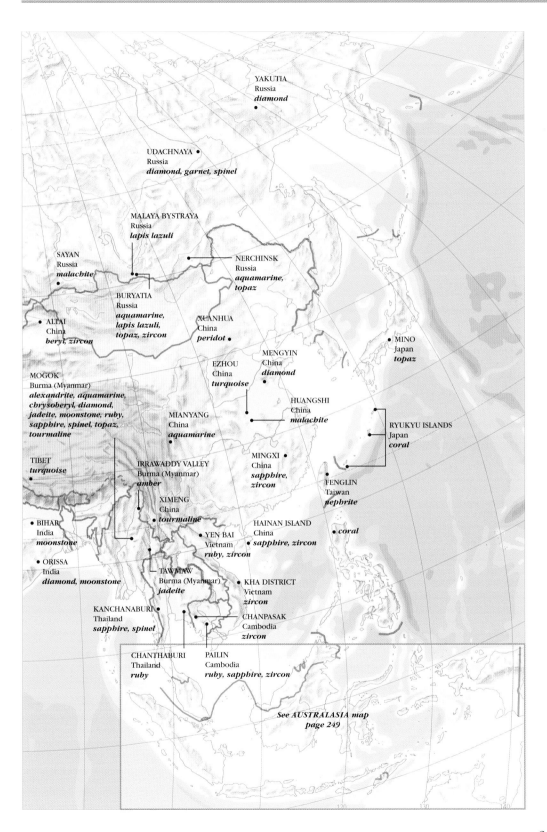

YAKUTIA
Russia
diamond

UDACHNAYA •
Russia
diamond, garnet, spinel

MALAYA BYSTRAYA
Russia
lapis lazuli

SAYAN
Russia
malachite

NERCHINSK
Russia
*aquamarine,
topaz*

BURYATIA
Russia
*aquamarine,
lapis lazuli,
topaz, zircon*

ALTAI
China
beryl, zircon

XUANHUA
China
peridot

MINO
Japan
topaz

EZHOU
China
turquoise

MENGYIN
China
diamond

MOGOK
Burma (Myanmar)
*alexandrite, aquamarine,
chrysoberyl, diamond,
jadeite, moonstone, ruby,
sapphire, spinel, topaz,
tourmaline*

MIANYANG
China
aquamarine

HUANGSHI
China
malachite

RYUKYU ISLANDS
Japan
coral

TIBET
turquoise

IRRAWADDY VALLEY
Burma (Myanmar)
amber

MINGXI
China
*sapphire,
zircon*

FENGLIN
Taiwan
nephrite

XIMENG
China
tourmaline

BIHAR
India
moonstone

HAINAN ISLAND
China
sapphire, zircon

• *coral*

YEN BAI
Vietnam
ruby, zircon

ORISSA
India
diamond, moonstone

TAWMAW
Burma (Myanmar)
jadeite

KHA DISTRICT
Vietnam
zircon

KANCHANABURI
Thailand
sapphire, spinel

CHANPASAK
Cambodia
zircon

CHANTHABURI
Thailand
ruby

PAILIN
Cambodia
ruby, sapphire, zircon

*See AUSTRALASIA map
page 249*

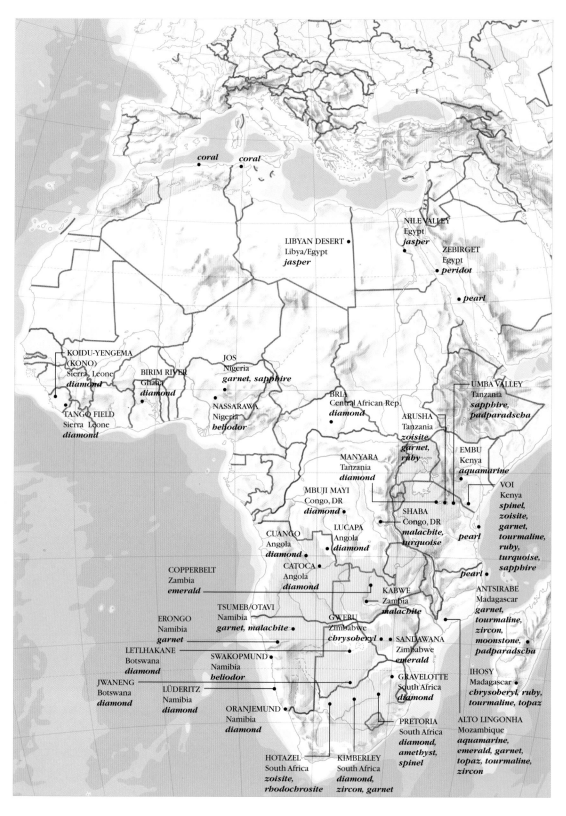

coral   coral

NILE VALLEY
Egypt
jasper

ZEBIRGET
Egypt
peridot

LIBYAN DESERT
Libya/Egypt
jasper

pearl

KOIDU-YENGEMA
(KONO)
Sierra Leone
diamond

BIRIM RIVER
Ghana
diamond

JOS
Nigeria
garnet, sapphire

UMBA VALLEY
Tanzania
*sapphire,
padparadscha*

TANGO FIELD
Sierra Leone
diamond

NASSARAWA
Nigeria
heliodor

BRIA
Central African Rep.
diamond

ARUSHA
Tanzania
*zoisite,
garnet,
ruby*

EMBU
Kenya
aquamarine

MANYARA
Tanzania
diamond

VOI
Kenya
*spinel,
zoisite,
garnet,
tourmaline,
ruby,
turquoise,
sapphire*

MBUJI MAYI
Congo, DR
diamond

SHABA
Congo, DR
*malachite,
turquoise*

pearl

CUANGO
Angola
diamond

LUCAPA
Angola
diamond

pearl

COPPERBELT
Zambia
emerald

CATOCA
Angola
diamond

KABWE
Zambia
malachite

ANTSIRABE
Madagascar
*garnet,
tourmaline,
zircon,
moonstone,
padparadscha*

TSUMEB/OTAVI
Namibia
garnet, malachite

GWERU
Zimbabwe
chrysoberyl

ERONGO
Namibia
garnet

SANDAWANA
Zimbabwe
emerald

LETLHAKANE
Botswana
diamond

SWAKOPMUND
Namibia
heliodor

IHOSY
Madagascar
*chrysoberyl, ruby,
tourmaline, topaz*

GRAVELOTTE
South Africa
diamond

JWANENG
Botswana
diamond

LÜDERITZ
Namibia
diamond

ORANJEMUND
Namibia
diamond

PRETORIA
South Africa
*diamond,
amethyst,
spinel*

ALTO LINGONHA
Mozambique
*aquamarine,
emerald, garnet,
topaz, tourmaline,
zircon*

HOTAZEL
South Africa
*zoisite,
rhodochrosite*

KIMBERLEY
South Africa
*diamond,
zircon, garnet*

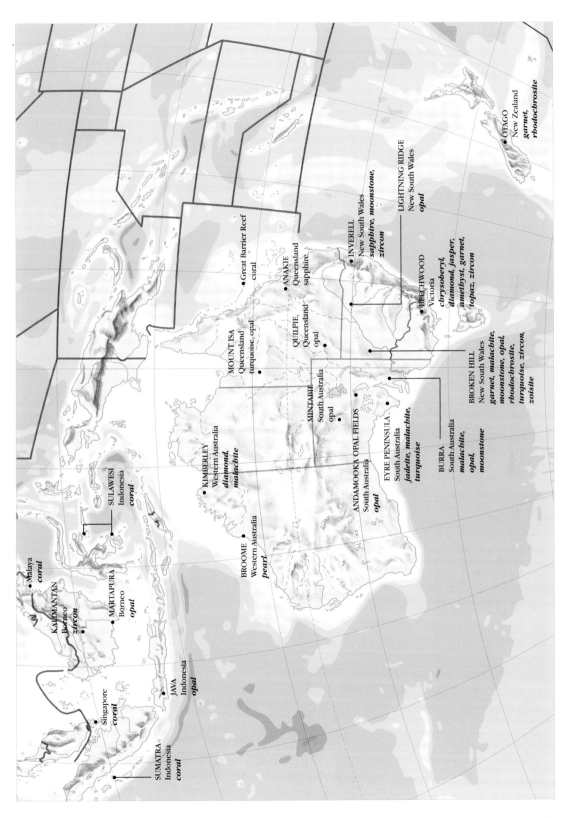

OTAGO
New Zealand
*garnet,
rhodochrosite*

LIGHTNING RIDGE
New South Wales
opal

INVERELL
New South Wales
*sapphire, moonstone,
zircon*

BEECHWOOD
Victoria
*chrysoberyl,
diamond, jasper,
amethyst, garnet,
topaz, zircon*

ANAKIE
Queensland
sapphire

• Great Barrier Reef
coral

QUILPIE,
Queensland
opal

MOUNT ISA
Queensland
turquoise, opal

BROKEN HILL
New South Wales
*garnet, malachite,
moonstone, opal,
rhodochrosite,
turquoise, zircon,
zoisite*

MINTABIE
South Australia
opal

BURRA
South Australia
*malachite,
opal,
moonstone*

EYRE PENINSULA
South Australia
*jadeite, malachite,
turquoise*

ANDAMOOKA OPAL FIELDS
South Australia
opal

KIMBERLEY
Western Australia
*diamond,
malachite*

BROOME
Western Australia
pearl

SULAWESI
Indonesia
coral

MARTAPURA
Borneo
opal

KALIMANTAN
Borneo
zircon

• Malaya
coral

Singapore
coral

JAVA
Indonesia
opal

SUMATRA
Indonesia
coral

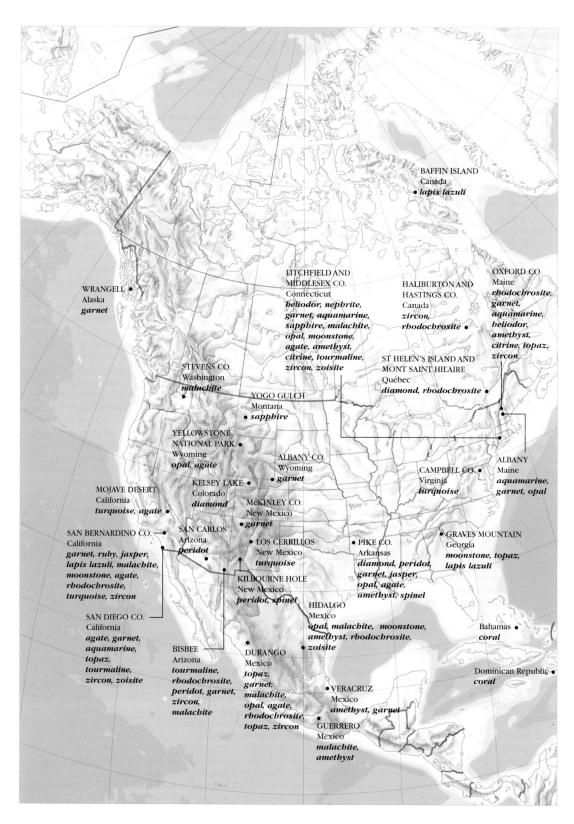

BAFFIN ISLAND
Canada
• *lapis lazuli*

WRANGELL •
Alaska
garnet

LITCHFIELD AND
MIDDLESEX CO.
Connecticut
*heliodor, nephrite,
garnet, aquamarine,
sapphire, malachite,
opal, moonstone,
agate, amethyst,
citrine, tourmaline,
zircon, zoisite*

HALIBURTON AND
HASTINGS CO.
Canada
*zircon,
rhodochrosite* •

OXFORD CO.
Maine
*rhodochrosite,
garnet,
aquamarine,
heliodor,
amethyst,
citrine, topaz,
zircon*

STEVENS CO.
Washington
malachite

YOGO GULCH
Montana
• *sapphire*

ST HELEN'S ISLAND AND
MONT SAINT HILAIRE
Québec
diamond, rhodochrosite •

YELLOWSTONE
NATIONAL PARK
Wyoming
opal, agate

ALBANY CO.
Wyoming
• *garnet*

CAMPBELL CO. •
Virginia
turquoise

ALBANY
Maine
*aquamarine,
garnet, opal*

MOJAVE DESERT
California
turquoise, agate •

KELSEY LAKE
Colorado
diamond

McKINLEY CO.
New Mexico
• *garnet*

SAN BERNARDINO CO. →
California
*garnet, ruby, jasper,
lapis lazuli, malachite,
moonstone, agate,
rhodochrosite,
turquoise, zircon*

SAN CARLOS
Arizona
peridot

• LOS CERRILLOS
New Mexico
turquoise

• PIKE CO.
Arkansas
*diamond, peridot,
garnet, jasper,
opal, agate,
amethyst, spinel*

• GRAVES MOUNTAIN
Georgia
*moonstone, topaz,
lapis lazuli*

KILBOURNE HOLE
New Mexico
peridot, spinel

HIDALGO
Mexico
*opal, malachite, moonstone,
amethyst, rhodochrosite,
zoisite*

SAN DIEGO CO.
California
*agate, garnet,
aquamarine,
topaz,
tourmaline,
zircon, zoisite*

BISBEE
Arizona
*tourmaline,
rhodochrosite,
peridot, garnet,
zircon,
malachite*

DURANGO
Mexico
*topaz,
garnet,
malachite,
opal, agate,
rhodochrosite,
topaz, zircon*

• VERACRUZ
Mexico
amethyst, garnet

Bahamas •
coral

Dominican Republic •
coral

• GUERRERO
Mexico
*malachite,
amethyst*

MUZO AND CHIVOR
Colombia
emerald

RIO GRANDE
DO NORTE
Brazil
*malachite,
opal, garnet,
zircon*

MARABA PARA
Brazil
amethyst

CEARA
Brazil
garnet

PAU D'ARCO PARA
Brazil
amethyst

PIAUI
Brazil
opal

BAHIA
Brazil
*emerald,
chrysoberyl,
alexandrite,
diamond,
malachite,
amethyst,
spinel,
tourmaline,
turquoise,
zircon*

GOIAS
Brazil
*malachite, topaz,
zircon*

CHUQUICAMATA
Chile
*tourmaline,
turquoise*

CATAMARCA
Argentina
*malachite, amethyst,
citrine, topaz*

COROMANDEL
Brazil
diamond

ESPIRITO SANTO
Brazil
*aquamarine,
chrysoberyl,
topaz*

OVALLE
Chile
lapis lazuli

CAPILLITAS
Argentina
turquoise

RIO GRANDE
DO SUL
Brazil
*amethyst,
citrine, agate*

ARAÇUAI
Brazil
*aquamarine,
heliodor,
alexandrite,
garnet,
tourmaline*

COQUIMBO
Chile
lapis lazuli

SALTO
Uruguay
agate

OURO PRETO
Brazil
*topaz,
diamond,
garnet,
tourmaline*

SAN LUIS PROVINCE
Argentina
turquoise

251

GLOSSARY

adamantine A "diamond-like" luster.

allochromatic Where the color is caused by small amounts of other elements or impurities not included in the chemical composition.

alluvial deposit Mud, silt or sand left by flowing water on flood plains, river beds, and estuaries.

amorphous Where there is no crystal structure and atoms are randomly arranged.

asterism A star-like effect caused by inclusions in a gemstone.

axis of symmetry The imaginary line that runs through a crystal and about which the crystal can be rotated while looking the same.

basal Cleavage on a horizontal plane by way of its base. Can sometimes be "peeled."

birefringence Difference between the maximum and minimum values of the refractive index.

cabochon Smooth domed gem, polished but unfaceted

chatoyancy Cat's-eye effect seen on some gemstones when cut as cabochons.

cleavage The direction along which the stone will break more easily.

conchoidal A shell-like fracture.

crown The upper part of the stone.

cryptocrystalline Where the crystal structure is too small to be seen with the eye.

crystalline Where the atoms are constituted in a regular and repeating three-dimensional structure.

crystallographic axis The imaginary line that runs through a crystal and indicates both the direction and length of the repeating pattern of an atom.

crystal symmetry Defined by the degree of regularity in the arrangement of atoms in the crystal structure.

crystal system There are seven crystal systems based on their crystal symmetry and crystallographic axis.

cubic system Crystal system with four three-fold axes of symmetry.

dichroic When a gem shows two different colors or shades of color when viewed from different directions.

dichroscope An instrument that reveals the pleochroism of gemstones.

dispersion Degree to which the spectral colors are reflected through the gem.

durability Specifically a gem's hardness, toughness, and stability.

extrusive Erupted volcanic igneous rocks are extrusive.

facet Flat, polished surface.

fracture A random, non-directional break that can be caused by impact, stress, or temperature change.

hardness A measure of how easily the surface of a gem can withstand abrasion.

hexagonal Crystal system with one vertical six-fold axis of symmetry.

hydrothermal veins Fluids that escape from magmas and may contain rare elements.

idiochromatic Where the color of a gemstone is caused by elements that are an essential part of the chemical composition.

igneous rock Derived from magma or lava that has solidified on or below the Earth's surface.

inclusion A gemstone may contain crystals of a different mineral within it.

intrusive Magma rocks formed without eruption.

isometric Crystal system with three axes, all of them equal in length and at 90° from the other.

isomorph Gems with a range of chemical compositions but the same crystal structure.

lapidary Person who cuts gems.

lattice structure The repeating pattern of the atoms.

loupe A hand lens.

luster The effect of the light reflected off the surface of the gem.

meleé A group of small diamonds each under 0.25 carats.

metamorphic rock Derived from magma or lava that has solidified on or below the Earth's surface.

microcrystalline Where the crystal structure is too small to be seen with the naked eye.

monoclinic Crystal habit with one two-fold axis of symmetry.

orthorhombic Crystal habit with either one two-fold axis of symmetry at the intersection of two mutually perpendicular planes, or three mutually perpendicular two-fold axes.

pavilion The bottom part of a gemstone.

pegmatite Intrusive igneous rock that produces a wider range of gemstones than any other rock type.

piezoelectric When a crystal is heated, compressed or vibrated and different electrical charges occur at opposite ends.

pinacoid Crystal habit comprised of only two parallel faces.

plane of symmetry The imaginary plane that divides a crystal such that the image on one side is the mirror image of the other side.

pleochroic A gemstone that appears different shades or colors from different positions.

polycrystalline Crystals composed of many crystal structures that have grown together.

polymorph A chemical composition that forms more than one crystal structure.

prismatic Crystal habit where a set of faces run parallel to an axis in the crystal.

pyramidal Where three planes intersect all three axes of the crystal.

refractive index The mathematical relationship between the angle at which light strikes a gemstone and the angle of refraction.

relative density (specific gravity) Difference in weight between a gemstone and water of equivalent volume.

schiller Relection of light by thin layers within the gemstone.

sedimentary rock Formed by accumulation and consolidation of minerals and organic fragments that have been deposited by water.

silk The effect from light reflected from patches of parallel inclusions.

stability Level of resistance to both chemical and physical alteration.

streak Fine layer of a softer mineral left on a gem.

tabular A crystal system where the crystals are flat in appearance, with two opposing sides being much wider than the other four.

tetragonal A crystal system with one vertical four-fold axis of symmetry.

toughness A measure of how well a stone can resist fracture.

tracer gem A gemstone that indicates a potentially rich mining area.

trichroic When a gem shows three different shades or colors when viewed from different directions.

triclinic Crystal system with either a center of symmetry or no symmetry.

twinning Where the crystal structure has parts that are reflected, repeated incorrectly, or rotated forming a twin crystal.

USEFUL ADDRESSES

American Gem Trade Association (AGTA)
P.O. Box 420643
Dallas, TX 75342-0643
USA
www.agta.org

American Museum of Natural History
Central Park West at 79th Street
New York, NY 10024-5192
USA
www.amnh.org

Natural History Museum
Mineralogy Department
Cromwell Road
London, SW7 5BD
England
www.nhm.ac.uk

Gemological Institute of America (GIA)
World Headquarters
The Robert Mouawad Campus
5345 Armada Drive
Carlsbad, CA 92008
USA
www.gia.org

Gemmological Association of Australia
P.O. Box A2175
Sydney South, NSW 1235
Australia
www.gem.org.au

Gemmological Association and Gem Testing Laboratory of Great Britain (GAGTL)
27 Grenville Street
London, EC1N 8TN
England
www.gagtl.ac.uk

International Colored Gems Association (ICA)
19 West 21st Street, Suite 705
New York, N.Y. 10010-6805
USA
www.gemstone.org

Mouawad USA, Inc.
45 West 45th Street, 15th Floor
New York, NY 10036
USA
www.mouawad.com

Multicolour Gems
www.multicolour.com

The Smithsonian Institution (Museum of Natural History)
10th Street and Constitution Ave., NW
Washington, D.C. 20560
USA
www.mnh.si.edu

ACKNOWLEDGEMENTS

The editors would like to give special acknowledgement to the following for their help in producing this book: Alan Jobbins; American Natural History Museum; Christie's; De Beers; House of Mouawad; ICA; Natural History Museum, London; NetComposite; Sotheby's; ThaiGem; Victoria and Albert Museum, London.

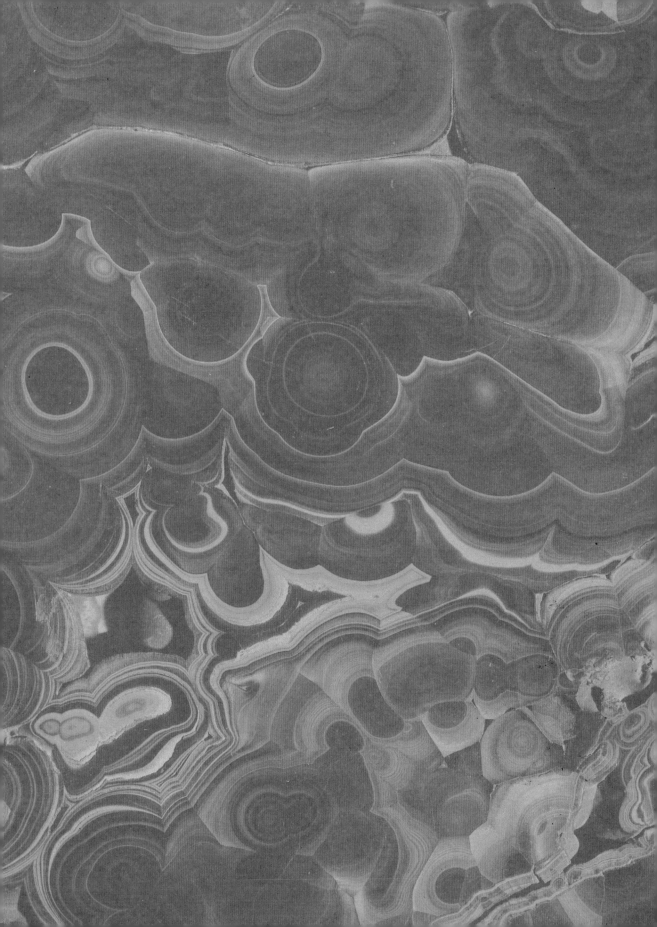